Statistics Applied With Excel

Franz Kronthaler

Statistics Applied With Excel

Data Analysis Is (Not) an Art

 Springer

Franz Kronthaler
University of Applied Sciences Grisons
Chur, Switzerland

ISBN 978-3-662-64318-1   ISBN 978-3-662-64319-8   (eBook)
https://doi.org/10.1007/978-3-662-64319-8

This book is a translation of the original German edition „Statistik angewandt mit Excel" by Kronthaler, Franz, published by Springer-Verlag GmbH Germany in 2021. The translation was done with the help of artificial intelligence (machine translation by the service DeepL.com). A subsequent human revision was done primarily in terms of content, so that the book will read stylistically differently from a conventional translation. Springer Nature works continuously to further the development of tools for the production of books and on the related technologies to support the authors.

This Springer imprint is published by the registered company Springer-Verlag GmbH, DE part of Springer Nature.
The registered company address is: Heidelberger Platz 3, 14197 Berlin, Germany

A Note to the Reader

It is often observed by lecturers that many students have great respect for the subject of statistics. Furthermore, although, many people had to learn statistics, they are nevertheless not much statistically literate, meaning that they typically do not think statistically when confronted with relationships and probabilities. One reason for this is probably that statistics is often taught in a heavily mathematical and, for many, somewhat incomprehensible way. Therefore, more than 5 years ago, I tried to write a simple, understandable stats book: a book that can be used across all subjects and that helps to apply statistics and trains in statistical thinking. The goal was simply to give everyone an easy and intuitive access to statistics. Feedback from colleagues and students indicate that the original idea and concept have worked well. The book, and I am very pleased about this, is now recommended and used at many universities and in various departments. Nevertheless, of course, one can always improve a book. This is what this new edition aims to do, without compromising the tried and tested concept. For example, in recent years, there has been an intense discussion in statistics about the concept of statistical significance. In addition, of course, there are always minor errors that need to be corrected.

But back to why statistics at all? One simple reason is that we are confronted with statistical numbers or correlations every day. If we open a newspaper, electronically or classically, we automatically read about correlations, and these have to be understood. For example, I recently read in a daily newspaper that young people drive cars that emit a lot of CO_2. The headline of the article was something like that, "When it comes to cars, young people don't care about the climate." This is about a correlation between age and the climate harmfulness of the car driven. Is this actually a correlation between age and climate-harmful cars? Or is there perhaps a third variable, the money available, behind this correlation? It's worth thinking about.

The ability to apply statistics and analyze data is also becoming increasingly important. This applies universally to all disciplines, businesses, public administrations, and people. If we observe the behavior of people, businesses, and other things, and are able to analyze those observations well, then we can make informed decisions. Hopefully, the better educated we are, the more informed our decisions are, and better is the world we live in. Ultimately, this is the motive why I started writing this book. It is to help analyze data, to apply statistics, and to think in statistical terms. It shows that data analysis is not as complicated as many believe. The book gives a simple approach to analyze a data set. At the same time, it enables you to better understand data analysis and statements by others. Who doesn't know the phrase "don't trust a statistic that you haven't falsified yourself"?

The special feature of this book is that it uses one data set to discuss the methods of statistics one by one. The data set can be found at www.statistik-kronthaler.ch. This makes it understandable how the methods of statistics build on each other and how gradually more and more information can be drawn out of a data set. The focus is on the basic contents of statistics that are needed to analyze a data set. Content that is rarely needed in data analysis is omitted. Thus, the book remains lean.

The second feature of the book is its focus on the application. The book is written in a non-mathematical way. I know from experience that mathematics often discourages readers from learning and applying statistics. For this reason, the book focuses on the concepts and ideas of statistics and how to apply them. I do not believe that statistics can be taught entirely without mathematics. However, it is possible to reduce the mathematics to its essentials, and to incorporate it in such a way that the focus is on the application of statistics rather than mathematics. The reader of the book should not even realize that mathematical concepts are being used to create knowledge. The idea is to have fun learning the benefits of using statistical methods.

The third feature is the easy substitutability of the data set used. It is possible to use a different data set and work through the book with it without problems. The analysis of a data set requires a systematic approach. This is represented by the structure of the book.

All three features together put you in a position to systematically analyze a data set without much effort.

Data analysis is fun!

New Features and Additions

Writing a textbook is always "work in progress," that is, you can always do something better or different or there are developments that should be considered. Therefore, it makes sense to rethink the contents and their presentation after some time. In addition, one always receives hints that can help to improve the book. So far, it appears that the original concept of the first edition has proven itself and therefore there is no need to revise the book conceptually. However, there is a lot of content that is worth integrating.

In statistics, there has been an intense discussion for some time about the term "statistical significance" and its use. Therefore, an important addition is to take this discussion into account and to include the resulting recommendations. In particular, effect size is added to the previous edition.

I thought for some time about whether to integrate price and quantity indices. In the end, because of the importance of the topic, for example, in inflation considerations, I decided to integrate this topic in the chapter on ratios.

A new chapter on analysis of variance has also been introduced. This allows the reader to test for group differences not only between two groups but also between several groups. The major advantage, however, is that this discusses an important concept in statistics, the concept of how to divide variance in an explained and an unexplained part.

A need of students is always the possibility to apply what they have learned directly with the help of applications and tasks. This is of course addressed in the new edition, and the number of applications is further increased. In particular, real data and data sources are integrated, so that the previous restriction to only simulated data is dissolved. In addition, students will learn how to find real data.

In addition, minor errors have been corrected and some concepts of statistics are discussed in more detail. For example, a recurring problem was the quartile calculation. With the way discussed so far, one receives slightly different values than those of the Excel output. This is now explained. In this context, it is also shown how boxplots can be drawn with Excel for several groups.

Lastly, the new edition is of course used to upgrade the book to the latest version of Excel, to Excel 2019. From this version onwards, it is also possible to draw map diagrams with Excel, which is of course shown.

Learning Features of the Book The book discusses statistics using a simulated real-life example. After reading the book, the reader should be able to systematically analyze a data set and extract the information contained in the data set. To ensure that this is achieved, the book uses additionally the following features.

Calculating by Hand and Calculating with Excel The examples are calculated both by hand (for a manageable number of observations) and with Excel (on a full data set). This makes it easier to understand the ideas and concepts of statistical methods, and at the same time we learn how to apply them to a larger data set using Excel. Excel can also be used to check tasks calculated by hand.

Freak Knowledge The freak knowledge section is designed to address knowledge and concepts that go beyond the content of the book. It addresses interesting aspects that generate additional knowledge without discussing them fully. The rubric is also intended to whet the reader's appetite for more.

Checkpoints At the end of each chapter, the most important points are briefly summarized in keywords. The reader thus gets an overview of what they should have learned in the chapter. In addition, they are challenged to think about these points again.

Applications and Solutions In addition to the checkpoints, exercises are provided at the end of each chapter to repeat the knowledge discussed. Understanding the concepts of statistics is encouraged by doing the calculations by hand, while practicing with Excel serves to further deepen the analysis of a data set. Solutions to the applications can be found at the end of the book. Also new in this book is the integration of tasks with real data. These data are provided in the applications in manageable numbers. However, you can and should always refer back to the original source and work with updated data. This provides access to real data.

Further Data Sets Also provided at the end of the book is another data set that allows the reader to apply what they have learned to a further example. In addition, instructors and students of other disciplines can find additional data sets at http://www.statistik-kronthaler. ch, which allow the reader to learn statistics using subject-related examples.

Acknowledgements

A project like this cannot be realized alone. It depends on support and inputs. With this, I would like to thank all those who have contributed to the success of the project. Some of them are especially worth mentioning here. First of all, these are my students. They bring the patience, sometimes they have to, to listen to me and support me with valuable questions and advice. But it is also my family, my wife and my children, who support me in word and deed. Thanks are also due in particular to the Springer team, Ms. Denkert, Mr. Heine, Ms. Herrmann, Ms. Ruhmann, and Ms. Stricker, who believed in the project from the outset and gave me the opportunity to publish the book. Ms. Ruhmann deserves special thanks for this new edition.

Contents

Part IV   Hypothesis Tests

List of Figures

List of Tables

Part I

Basic Knowledge and Tools to Apply Statistics

Statistics Is Fun

1

Statistics is fun and interesting. Interesting because we use statistics to discover relationships and understand how people and also other things, such as companies, behave. Fun, simply because it is fun to discover things. At the same time, statistics is important. If we can observe and evaluate how people, companies, or other objects behave, then we simply understand many things better. In this chapter we clarify why statistics is interesting and relevant, at the same time we clarify important terms.

1.1 Why Statistics?

Why statistics? This question is asked by many students of business administration, information science, psychology, sociology, tourism, etc. Let statisticians deal with statistics. Why should we learn statistics? If this question is answered, then we have already won. We are motivated to deal with statistics and statistics is already half learned. However, statistics is important to all of us for a number of good reasons.

First, a good portion of the knowledge in the world is based on data and data analysis. We can divide knowledge into three types, experience knowledge, theoretical knowledge, and empirical knowledge. Experience knowledge is knowledge that we acquire throughout life based on our own experiences, for example: on a hot stove I burn my fingers! If we think about this for a moment, we immediately realize that the smallest part of our knowledge is based on such experiences. Theoretical knowledge is, loosely speaking, knowledge that is logically derived from observations. This already brings us to the data. Theoretical knowledge is built on observations. But it does not stop there. It is verified and improved with the help of new observations and data, which brings us to empirical knowledge. For example, not so long ago it was generally accepted knowledge that the number of children born per woman is lower in richer countries than in poorer countries.

© Springer-Verlag GmbH Germany, part of Springer Nature 2023
F. Kronthaler, *Statistics Applied With Excel*,
https://doi.org/10.1007/978-3-662-64319-8_1

This knowledge arose from the observation that birth rates decline as countries become richer. More recent observations show that this is still true today, but with the small difference that above a certain level of wealth, the number of children per woman increase again.

The second argument is simply that data is sometimes used to make opinions, at worst to manipulate individuals. If we understand how knowledge or supposed knowledge is generated from data, then we are better able to detect attempts of manipulation.

And finally, data analysis helps us to make good decisions. Hal Varian, chief economist at Google, said some time ago the following: "I keep saying the sexy job in the next ten years will be statisticians. [...] The ability to take data - to be able to understand it, to process it, to extract value from it, to visualize it, to communicate it - that's going to be a hugely important skill in the next decades [...]." Companies like Google, but also other big and small companies, process data to get to know their customers and employees, to enter new markets or to check the image of their products.

There are plenty of good reasons to get into touch with statistics. But the best one is that learning the techniques of data analysis is not as complicated as some might fear. All it takes is a lecturer who explains the concepts of statistics simply, "a lecturer on your side".

1.2   Checkpoints

- *Knowledge is based on experience, theories, and data, but most of our knowledge is based on data.*
- *Data is sometimes used to manipulate.*
- *Sound data analysis helps to make good decisions.*

1.3   Data

In statistics, we are concerned with the analysis of data. With that, we first need to clarify what data is and how a data set is built up. A data set is a table which contains information about people, companies or other objects. There are three pieces of information in a data set: (1) for whom do we have information, (2) about what do we have information, and (3) the information itself. Let's imagine, we are studying ten people and we want to know their sex, age, income, and also how much they like eating cheeseburgers. We can display this information in a table.

Table 1.1 shows a first data set. This data set is structured as follows: The first column (column A) shows about whom or which objects we do have information. No. 1 can stand for Mr. Tchaikovsky, no. 2 for Mrs. Segantini, no. 3 for Mrs. Müller, and so on. Column 1, technically speaking, informs about the objects of the study.

In the first row (row 1) we can read about what we do have information. In this data set, there is information about sex, age, income, and how much the individuals

Table 1.1 Data set persons, sex, age, income, burger

	A	B	C	D	E	F	G	H
1	person	sex	age	income	cheeseburger			
2	1	male	55	10000	very much			
3	2	female	23	2800	very much			
4	3	female	16	800	not really			
5	4	male	45	7000	not at all			
6	5	male	33	8500	somewhat			
7	6	female	65	15000	not really			
8	7	male	80	2000	not really			
9	8	female	14	0	not at all			
10	9	female	30	5200	very much			
11	10	female	45	9500	not at all			

like to eat cheeseburgers. In a statistical sense, these are the variables. They describe the characteristics of the objects of the study.

The cells from column A and row 2 (A2 for short) to column E and row 11 (E11) give the information about each person and variable. Person no. 1, Mr. Tchaikovsky, is male, 55 years old, has an income of 10,000 Swiss francs, and loves to eat cheeseburgers. Person no. 10 is female, 45 years old, has an income of 9500 francs, and does not like eating cheeseburgers. This way we can read each line and get information about each person. But we can calculate how people behave with regard to the individual variables too. For example, we can read relatively quickly that there are 4 men and 6 women in the data set, and that 3 out of 10 people (30% of the study objects) like cheeseburgers a lot, and so on.

A data set is usually organized in the way shown above. The rows contain the objects of the study. The columns contain the variables. The intersections between columns and rows provide the information about the objects of the study.

In addition, there is another detail that we need to know about. Usually, a data set contains only numbers next to the variable names, no other words. The reason is simple, numbers are easier to calculate with. Exceptions are of course possible, e.g. if we intentionally collect words or texts and want or need to continue working with them. For example, in many questionnaires there is a question where respondents can provide further information.

Table 1.2 Coded data set persons, sex, age, income, burger

A	B	C	D	E
person	sex	age	income	cheeseburger
1	0	55	10000	1
2	1	23	2800	1
3	1	16	800	3
4	0	45	7000	4
5	0	33	8500	2
6	1	65	15000	3
7	0	80	2000	3
8	1	14	0	4
9	1	30	5200	1
10	1	45	9500	4

Table 1.2 shows the same data set, except that now all information is given in numbers. We have converted the information that was present as words into numbers. Technically speaking, we have coded the verbal information, i.e. assigned with numbers.

Coding requires to create a legend in which we note the meaning of the numbers. The legend should contain the following information: variable name, variable description, values, missing values, scale and source of the data.

Table 1.3 displays the legend for the data set. The variable name is the short name used for the variable. The variable description is used to precisely describe the meaning of the variable. The idea is simply that later, perhaps a year from now, we will still need to know what, for example, the variable cheeseburger actually means. The values column tells us how the data is defined, for example, that for the variable sex, 0 represents a man and 1 represents a woman. Missing values are denoted in this legend by the letters n.d., where n.d. stands for not defined. We do not need to define missing values because our data set contains information on all individuals and all variables. However, it is often the case that information is not available for all persons for all variables. We then can help ourselves by either leaving the cell in the data set empty or by filling it with a defined value that cannot occur for the corresponding variables, usually 8, 88, 888, ..., 9, 99, 999, ... The defined value can be supplemented by a note indicating why the value is missing, e.g. 8, 88, 888, ..., can stand for "no answer possible" or 9, 99, 999, ..., for "answer refused". In the first case, the respondent was unable to give an answer. In the second case, the respondent refused

Table 1.3 Legend for the data set persons, sex, age, income, burger

	A	B	C	D	E
1	variable name	variable description	values	missing values	scale
2	person	index number of surveyed persons			
3	sex	sex of surveyed persons	0=male 1=female	n.d.	nominal
4	age	age of surveyed persons	in years	n.d.	metric
5	income	income of surveyed persons per month	in CHF	n.d.	metric
6	cheeseburger	How do the people surveyed like to eat cheeseburgers?	1=very much 2=somewhat 3=not really 4=not at all	n.d.	ordinal
7	Source: Own survey 2021.				

to answer for whatever reason. The scale column informs about the scale of measurement of the data. This information is of huge relevance when selecting statistical methods. We will discuss this in the next section. In addition, we need the information about the source of the data. This information helps to assess the trustworthiness of the data. In our case, we have collected the data ourselves, the source is our own survey from 2021. If we had worked well, we know that the data set has a good quality. But if we know the data is from a less trustworthy source, we don't have to throw it away because of that. We do, however, need to take the uncertainty about the data quality into account when analyzing and interpreting the data.

You may also have noticed in Table 1.3 that in the second row only the variable name and the description of the variable contain an information. The cells for values, missing values and scale are empty. Here we do deal with the running index of the observed persons, accordingly the values are a consecutive number. We do not have any missing values, since all observed persons are in the data set and we do not need a scale, since this column of the data set is not used for calculations.

1.4 Checkpoints

- *A data set is organized into columns and rows. The rows contain the objects of the study, and the columns contain the variables.*

- *A data set is usually organized using numbers. It is easier to calculate this way.*
- *The legend of the data set should contain the variable name, the variable description, the definition of the values, the information about missing values and the scale.*

1.5   Scales: Lifelong Important in Data Analysis

We are getting closer to data analysis with huge steps. Before we begin, we need to address the scale of measurement of variables. The scale of the data essentially determines which statistical methods are applicable. We will see this in each of the following chapters, and will continue to see it throughout the rest of our statistical lives. It is worth spending time here memorizing the types of scales and their properties.

We have already learned about the different types of scales in our legend. Data can be nominal, ordinal, or metric.

Nominal means that the variable only allows to distinguish between the persons and objects of the study. In our example, the variable sex distinguishes between a man and a woman, a man is a man, a woman is a woman, and a man is not a woman. Technically speaking, in this case $0 = 0$, $1 = 1$ and $0 \neq 1$. Other examples is religion with the characteristics Christian, Buddhist, Muslim, Jewish etc., the corporate form of companies with the characteristics limited company, public limited company, ..., the blood type of persons or the breed of dogs. In these examples we also see that there are nominal variables with two or more than two characteristics. In the case of a nominal variable with two characteristics, e.g. the variable sex with male and female, we also speak of a dichotomous variable or a dummy variable.

In the case of an ordinal variable, a ranking adds to the distinction, for example in the form of better-worse or friendlier-unfriendlier. In our example, the variable cheeseburger indicates how much people like to eat cheeseburgers. We have a ranking from very much, somewhat, not really to not at all. We know whether a person likes cheeseburgers very much, somewhat, not really or not at all. However, we do not know the distance between the characteristics. Technically speaking, in our example, $1 = 1$, $2 = 2$, $3 = 3$, $4 = 4$, and $1 > 2 > 3 > 4$. The distances between the values 1, 2, 3, and 4 are not defined. A classic example of this are school grades. With school grades, we only know whether we have achieved an excellent, very good, good, satisfactory, sufficient, unsatisfactory or very poor grade. The extra learning we need to get from a sufficient grade to a satisfactory grade or from a good grade to a very good grade varies from grade to grade, from teacher to teacher, and from exam to exam. Further examples for ordinal variables are satisfaction with a product with the characteristics very satisfied, satisfied, less satisfied, dissatisfied or the injury sports risk with the expressions high, medium, low.

For metric variables, in addition to the distinction and ranking, the distance between two expressions is also defined. In our example, age and income are metric. A person with 21 years is older than one with 20 years and the difference between both values is exactly 1 year. Here, we know exactly how much time passes from the 20th birthday to

Table 1.4 Scales of measurement

Scales	Characteristics	Examples
Nominal	Distinction $0 = 0$, $1 = 1$, and $0 \neq 1$	Sex: male, female; Hair color: black, brown, red, ...; Blood group: A, B, AB and 0
Ordinal	Distinction $1 = 1$, $2 = 2$, $3 = 3$, $4 = 4$ and $1 \neq 2$, ... Rank order $1 > 2 > 3 > 4$	School grades: excellent, very good, good, satisfactory, ...; Rating: very much, somewhat, not much, not at all
Metric	Distinction $X = X$, $Y = Y$, $Z = Z$ and $X \neq Y$,... Rank order $X > Y > Z$ Distance is defined $X - Y$ resp. $Y - Z$	Age: 5 years, 10 years, 20 years, ...; Pocket money: CHF 10, CHF 20, CHF 25, ...; Weight: 50 kg, 65 kg, 67 kg, ...

the 21st birthday or from the 35th birthday to the 36th birthday. Technically speaking, $X > Y > Z$ and $X - Y$ or $Y - Z$ is clearly defined. Furthermore, we have theoretically infinite intermediate values between two expressions. Let's look again at the example from the 20th to the 21st year of life. In between there are 365 days or 8'760 h or 525'600 min, etc. Other examples of metric data are temperature, turnover, distance between two places, etc. (Table 1.4).

In applied data analysis, the question arises again and again whether we can treat ordinal data the same way as metric data. We will be able to grasp this later. This question can be answered with not really, but ... . Actually, metric equals metric and ordinal equals ordinal. Nevertheless, ordinal data are often treated like metric data. This is justifiable if an ordinal variable has enough characteristics, about seven (e.g., from 1 to 7), more is better, and the variable is normally distributed. We will learn later in this book about the normal distribution. More elegant is to create a quasi-metric variable. Instead of asking, "How do you like eating cheeseburgers?" and having that question answered on a scale from 1 to 7, we can ask the same question and let instead put a cross on a line.

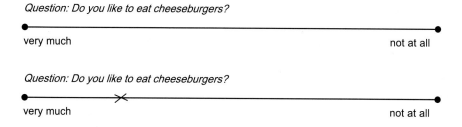

Afterwards we measure the distance between the beginning of the line and the cross as accurately as possible. In this case a small distance means that someone likes cheeseburgers very much. The closer the distance gets to the maximum possible distance, the less the person likes to eat cheeseburgers. If we try this a few times, simulating for several people, we quickly notice that we get a lot of different values, creating a quasi-metric variable.

As long as we are analyzing data, scales of measurement will never leave us alone!

1.6   Checkpoints

- *Nominal variables allow for a distinction between the objects of study.*
- *Ordinal variables include a ranking in addition to the distinction.*
- *Metric variables allow in addition to distinction and ranking a precise distance measurement.*
- *Ordinal data can be treated like metric data, if enough characteristics are possible and the data is normally distributed. More elegant is the creation of a quasi-metric variable.*
- *The scale level plays a decisive role in determining which statistical methods can be used.*

1.7   Software: Excel, SPSS, or "R"

Fortunately, data analysis is no longer done by hand. We have several computer programs to assist us, such as Excel, SPSS, or "R". Which of these programs should we use then?

Excel is Microsoft's spreadsheet program. It is not a purely statistical software but has a variety of functions that can be used to store, organize and analyze data. The advantages of Excel are that it can be operated through a graphical interface, we usually know it, and it is typically installed on our personal computer. Moreover, it is usually available in the company in which we work. The disadvantage is that not all statistical applications are available.

The statistical program SPSS "Statistical Package for the Social Sciences" is a professional statistical software. Accordingly, it has most of the well-known applications of statistics. Like Excel, SPSS works through a graphical interface and programming skills are not necessarily required. The major disadvantage of SPSS is its price. Moreover, in a company where SPSS is not present, we cannot easily have or use it.

A free and also professional statistical software is available with the program "R". "R" is an open source software, and a lot of people are involved in its development. As SPSS, it allows the application of most of the known methods of statistics. The big advantage of "R" is its free availability, the large scope of statistical methods and the increasingly wide masses that use this program and offer help on the internet. However, a disadvantage is that the use of "R" partly requires programming skills. This is especially the case with more complex statistical applications.

Which software should we use? Excel is for several reasons a good solution. We already know Excel, it is available on our computer and on the company computer, and Excel has a variety of statistical methods. In addition, many organizations provide data as Excel files.

If you would like to do data analysis more professional later, you should consider switching to "R" at some point. The switch is easy to make if you already know the basics of statistics. Information about "R" can be found at https://www.r-project.org. Furthermore, there is a lot of literature that deals with "R". Among others, a colleague and I wrote a simple introduction to "R".

1.8 Case Studies: The Best Way to Learn

Data analysis is best learned by using a hands-on example. In the book we use a central data set for this purpose. The data set used contains 10 variables and 100 observations. It is designed to be as close to reality as possible, while covering most of the issues that arise when analyzing a data set. The data set is available on the internet platform www.statistik-kronthaler.ch.

By using the central data set throughout the book, we see how the methods of statistics build up on each other and can be used to systematically analyze a data set. At one point or another in the book, the data set will still be not sufficient and we will use other data sets too. These data sets are also available at www.statistik-kronthaler.ch.

1.9 Case Study: Growth of Young Enterprises

Let us imagine that we have to write a first paper in our studies, a bachelor thesis, a master thesis or our doctoral dissertation about growth of young enterprises. For this purpose, we survey young enterprises and now we are faced with the task of analyzing the data. Specifically, we have the following data set with information about young enterprises and their founders (Table 1.5).

In the table, the data available for the first twelve enterprises surveyed are displayed. We use this data in the book when we calculate by hand. When we analyze the data using Excel, we refer to the complete data set with a total of 100 companies. Larger data sets of 100 or more observed objects are the norm in practice. But we will also occasionally have data sets with fewer observed persons or objects. Now still the legend is missing, without it we are not able to work with the data (Table 1.6).

Table 1.5 Data set young enterprises

	A	B	C	D	E	F	G	H	I	J	K	L
1	enterprise	growth	expectation	marketing	innovation	sector	motive	age	experience	self-assessment	education	sex
2	1	5	2	29	3	0	3	32	7	5	1	0
3	2	8	1	30	5	1	2	26	8	4	2	1
4	3	18	3	16	6	0	2	36	10	5	1	0
5	4	10	2	22	5	1	1	43	7	3	3	1
6	5	7	1	9	9	1	2	43	6	3	1	1
7	6	12	2	14	8	1	3	48	10	4	4	0
8	7	16	2	26	0	1	2	19	11	5	3	0
9	8	2	1	26	4	1	1	33	4	2	1	0
10	9	4	1	28	8	1	2	42	0	1	4	1
11	10	10	1	31	0	1	3	27	4	2	4	1
12	11	7	1	6	3	1	2	30	8	4	1	1
13	12	9	1	30	0	1	2	23	12	5	1	1

1.10   Applications

1.1 Give three reasons why statistics is useful.

1.2 What information does a data set contain?

1.3 What scales are important for statistical data analysis and what properties do they have?

1.4 What are the scales of the variables sector (industry, service), self-assessment (very experienced, experienced, less experienced, not experienced), and turnover (in Swiss francs)?

1.5 What is the scale of the variable education (secondary school, A-level, bachelor, master)?

1.6 Variables can sometimes be measured using different scales. Consider how the educational level of individuals can be measured in a metric way.

1.7 The hotel satisfaction of guests should be measured in a quasi-metric way. For this purpose, create eight lines with a length of 10 cm, measure a hypothetical value for each guest and create a small data set with the variables guest, age, gender and hotel satisfaction.

Table 1.6 Legend for the data set young enterprises

	A	B	C	D	E
1	variable name	variable description	values	missing values	scale
2	enterprise	index number of surveyed enterprises			
3	growth	average growth rate of turnover in the last five years	in percent	n.d.	metric
4	expectation	expectation about the future development of the enterprise	1=better than before 2=no change 3=worse than before	n.d.	ordinal
5	marketing	expenditure on marketing in the last five years	in percent of turnover	n.d.	metric
6	innovation	expenditure on innovation in the last five years	in percent of turnover	n.d.	metric
7	sector	sector in which the company operates	0=industry 1=service	n.d.	nominal
8	motive	founder motive of the founder	1=unemployment 2=implement idea 3=higher income	n.d.	nominal
9	age	age of the company founders at the time of foundation	in years	n.d.	metric
10	experience	industry professional experience of the company founders	in years	n.d.	metric
11	self-assessment	self-assessment of the industry professional experience of the company founders	1=very experienced 2=experienced 3=neither nor 4=less experienced 5=not experienced	n.d.	ordinal
12	education	highest school degree of the company founder	1=secondary school 2=high school 3=university of applied sciences 4=university	n.d.	ordinal
13	sex	sex of the company founder	0=male 1=female	n.d.	nominal
14	Source: Own survey 2021.				
15	© Kronthaler, 2021				

1.8 Why are scales important for data analysis?

1.9 Why do we need a legend for a data set?

1.10 Open our data set data_growth.xlsx with Excel and learn about it. Answer the following questions: What is the number of observations? How many metric, ordinal, and nominal variables are in the data set? How are the metric, ordinal, and nominal data measured?

1.11 How trustworthy is our data set data_growth.xlsx?

1.12 Name one trustworthy data source each for your country, continent and the world?

1.13 Think of three questions that are likely to be analyzed with data. Find the appropriate data sources.

Excel: A Brief Introduction to the Statistical Tools

2

Before we start with the chapter, I would like to give a few minor notes about it. First, the chapter gives a brief introduction to the statistical tools of Excel 2019, it does not discuss all the possibilities, which Excel provides. For those who want to delve more deeply into Excel 2019, I recommend one of the specialized manuals available. Secondly, the tools available in Excel 2019 are not much different from the tools of past versions of Excel. So, it is not a problem if we are working with an older Excel version. Third, those who are familiar with Excel can safely skip the chapter. Last, all calculations in the book were done either by hand or with Excel 2019, and all figures were created with it as well. This gives a first hint of the possible use of Excel in statistics. We will learn now a little bit more.

To use Excel for data analysis, first we have to open an Excel worksheet. Figure 2.1 displays an Excel worksheet and at a glance the things we should be familiar with. At the top of the worksheet is the name of the Excel file. Below that is the ribbon with the tabs. Here we find the commands we can use for our data analysis. There is also a tab called Help. We click on this tab when we need help with Excel. With the question mark and a keyword, we usually get hints on how to solve the problem.

We use the tab File to manage our Excel file. We can use it, for example, to save the file after calculations or data typed in. In addition to the "Save" command, we also have the "Save As" command available here. We use "Save As" when we would like to save the worksheet but keep the original data set unchanged. It is always advisable to keep the original data set, as it could be the case that an error occurs during saving or analysis and we want to start all over again.

The Insert tab is used when we want to take advantage of Excel's various options to create graphs and tables, for example, to create a bar chart, a pie chart, a scatter plot, or even a pivot table. During working through the book, we'll learn about these options. Perhaps a note here, even today I usually create the graphs I use for publications using Excel.

© Springer-Verlag GmbH Germany, part of Springer Nature 2023
F. Kronthaler, *Statistics Applied With Excel*,
https://doi.org/10.1007/978-3-662-64319-8_2

In particular, in the Formulas tab, we find the Insert Function command. We will use this command again and again throughout the book when we use the various statistical functions, for example, to calculate a mean, a correlation coefficient, or something else.

The Data tab gives access to special statistical functions available in Excel. We will find these statistical functions in the "Data Analysis" command. The "Data Analysis" command is not automatically available. We need to activate the Data Analysis tool first. How, we will discuss in a moment. We also use the tab to sort data, for example, by size in ascending or descending order. We will need the sort function on various occasions when analyzing data with Excel.

The Review tab is in particular used to avoid accidentally overwriting the data. All the data in the internet available for the book is protected with the sheet protection function. This helps that the data cannot be easily overwritten. However, the sheet protection can be easily removed by clicking on the "Unprotect Sheet" command. The sheet protection can also be password protected.

In the cells we write our information as described in Chap. 1. To do this, we click with the left mouse button on a cell. The cell is then outlined in green and we can enter numbers or even text. To move from one cell to the next, we use either the mouse or the arrow keys on the keyboard.

Two more notes. Sometimes we get visual hints in the cells, such as a small triangle at the top left of the cell. This indicates that there are problems in the spreadsheet, for example with the calculation or the data format. The triangle is backed with information that gives information about the existing problem. At the bottom we see the worksheets

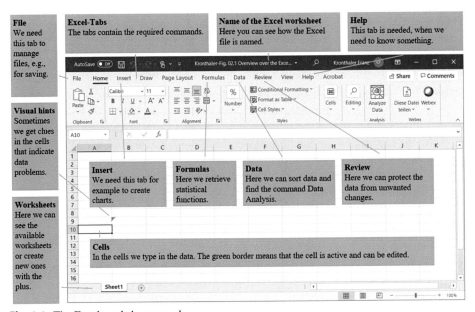

Fig. 2.1 The Excel worksheet at a glance

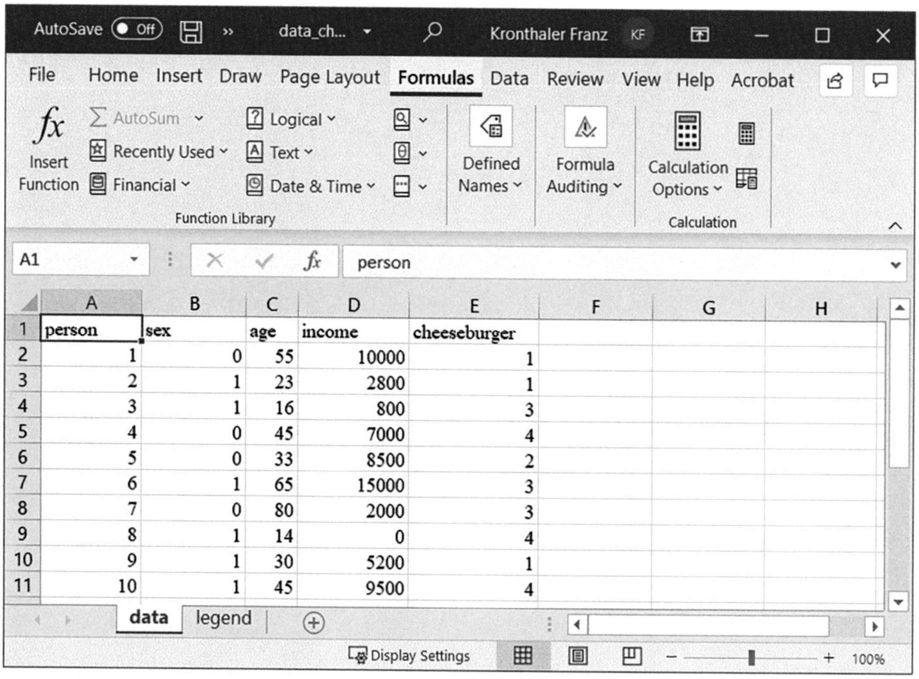

Fig. 2.2 The formulas tab in Excel

active. In our case this is usually the worksheet with the data and the worksheet with the legend (compare Chap. 1). If necessary, we can add further worksheets with the plus, e.g. with the results of our calculations.

Now, before we start to analyze data, let's look at the two ways Excel offers to perform statistical procedures. As mentioned earlier, the first way is to use the command Insert Function from the Formulas tab (Fig. 2.2).

If we click on the Insert Function button, a window opens with the available functions (Fig. 2.3). We select the category Statistical to have access to the statistical functions.

The second way is to use Excel's Data Analysis Tool. First we have to activate it. To do this, we click on the File tab and then on Options (Fig. 2.4).

In the window that appears, we click on Add-Ins. There we select the line Analysis Functions and then click on Go (Fig. 2.5).

In the next window, we choose the Analysis ToolPak and click on OK (Fig. 2.6).

Now is in the Data tab, the Data Analysis command available, look for it. If we click on this command, the Data Analysis Tool of Excel opens and its statistical functions are available (Fig. 2.7).

We will use these functions one by one. Let us finally start with data analysis.

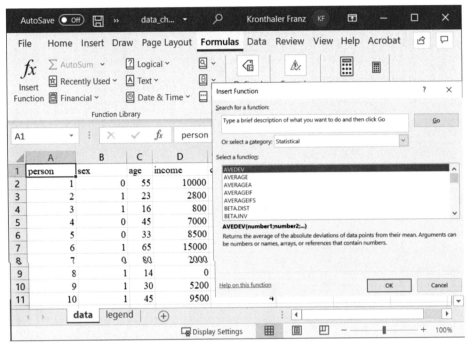

Fig. 2.3 The insert function window in Excel

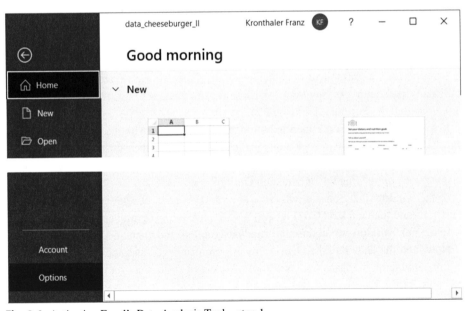

Fig. 2.4 Activating Excel's Data Analysis Tool—step 1

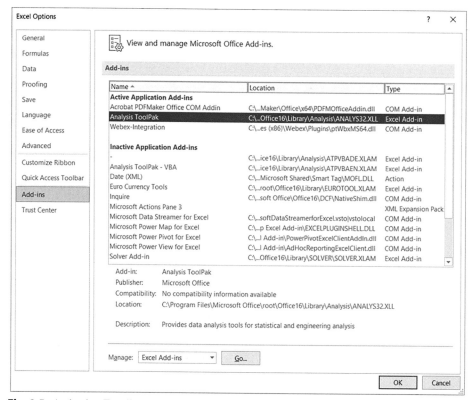

Fig. 2.5 Activating Excel's Data Analysis Tool—step 2

Fig. 2.6 Activating Excel's Data Analysis Tool—step 3

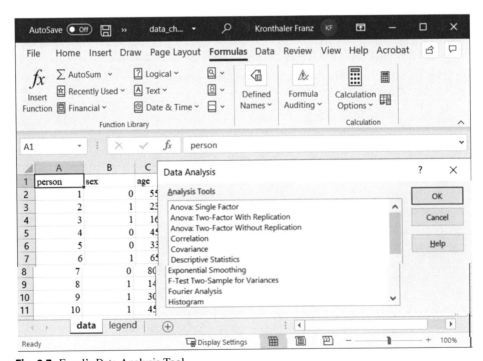

Fig. 2.7 Excel's Data Analysis Tool

Describing People or Objects, or Simply Descriptive Statistics

In descriptive statistics, we report on people and objects that we observe. For example, as customers of McDonalds or Burger King, we want to know how much meat is in the cheeseburgers we buy. In such a case, we observe the cheeseburgers, that is, we weigh the meat in each cheeseburger, then we collect the information and evaluate it for all the observed cheeseburgers.

In our case study of young enterprises, we can use descriptive statistics to obtain information about the surveyed firms. After the analysis, we know how the "observed firms" behave. We have several means to describe the observations. In Chap. 3 we deal with the average, in Chap. 4 with the deviation from the average and in Chap. 5 we ask how groups of observations behave and how we can graphically display the data. Subsequently, we consider in Chap. 6 relationships between variables and in Chap. 7 we show how new knowledge can be generated from existing data.

Average Values: How People and Objects Behave in General

3

3.1 Average Values: For What Do We Need Them

We use average values to analyze how people and objects behave in general, e.g. how much meat cheeseburgers contain on average. For this purpose we have different average values at our disposal: the arithmetic mean, the median, the mode and the geometric mean. These values are also called measures of central tendency. We'll look at them and their application in a moment. Finally, this leads to data analysis.

3.2 The (Arithmetic) Mean

The arithmetic mean, often just called the mean, is the measure of central tendency that is most commonly used. It is simple to calculate. We add up all the values of a variable and then divide by the number of observations:

$$\bar{x} = \frac{\sum x_i}{n}$$

$\sum$ is the Greek symbol for sum. Whenever we see this symbol, we add up values,
$\bar{x}$ is the (arithmetic) mean,
x_i are the observed values and
n is the number of observations.

Importantly, the arithmetic mean assumes metric data, i.e., we should calculate it only if our scale level is metric.

Let's take the case study growth of young enterprises and recall the data (see Table 1.5). We want to know how the first six companies grow on average. We find the data for this

© Springer-Verlag GmbH Germany, part of Springer Nature 2023
F. Kronthaler, *Statistics Applied With Excel*,
https://doi.org/10.1007/978-3-662-64319-8_3

by the variable growth. The growth rates are:

$$x_1 = 5$$
$$x_2 = 8$$
$$x_3 = 18$$
$$x_4 = 10$$
$$x_5 = 7$$
$$x_6 = 12$$

The sum of the growth rates equals:

$$\sum x_i = x_1 + x_2 + x_3 + x_4 + x_5 + x_6 = 5 + 8 + 18 + 10 + 7 + 12 = 60$$

Now, we divide the sum by our six companies and we get the average growth rate:

$$\bar{x} = \frac{\sum x_i}{n} = \frac{60}{6} = 10$$

Since the growth rate was measured in percent (see Table 1.6), the average growth rate is 10%. Hence, our six companies have grown at an average rate of 10% over the last 5 years. For example, we are also interested in, how old our founders are on average. This can be calculated accordingly as follows:

$$\bar{x} = \frac{\sum x_i}{n} = \frac{32 + 26 + 36 + 43 + 43 + 48}{6} = 38$$

Our founders are on average 38 years old. They spend an average of 20% of their revenue on marketing and another 6% of their revenue on innovation. Furthermore, the founders have an average of 8 years of industry professional experience.

Bingo, so we already know quite a bit about our enterprises and their founders.

Freak Knowledge
We denote the arithmetic mean with $\bar{x}$ and the number of observations with the lower case letter n. Strictly speaking, we denote it this way only when analyzing a sample. Here we do not analyze all objects or persons of interest, but only a part of it. If we have observed the population, i.e., all persons or objects of interest, the arithmetic mean is denoted by the Greek letter μ, and the number of observations by the upper case letter N.

Table 3.1 Growth rates in 5% steps

Class	Growth rate from over ... to ... [in percent]	Number of enterprises
1	−10 to −5	3
2	−5 to 0	8
3	0 to 5	27
4	5 to 10	34
5	10 to 15	23
6	15 to 20	4
7	20 to 25	1

Sometimes it is the case that the data have been already been processed and are only available in categories. For example, the growth rates of our 100 firms might be available as follows (Table 3.1).

In such a case, we have the growth rates only for groups of enterprises, not for individual firms. If we still want to calculate the arithmetic mean, we use the following formula:

$$\bar{x} \approx \frac{\sum \bar{x}_j \times n_j}{n}$$

$\bar{x}_j$ is the mean value of the respective classes,
n_j is the number of enterprises in the respective class and
n is the total number of observed firms.

For class 1 with three firms, the mean of the growth rate is −7.5, for class 2 with eight firms −2.5, for class 3 with 27 firms 2.5, and so on. Thus, the mean can be calculated as follows:

$$\bar{x} \approx \frac{\sum \bar{x}_j \times n_j}{n}$$

$$\approx \frac{-7.5 \times 3 - 2.5 \times 8 + 2.5 \times 27 + 7.5 \times 34 + 12.5 \times 23 + 17.5 \times 4 + 22.5 \times 1}{100} \approx 6.6$$

The average growth rate of firms over the past 5 years is approximately 6.6%. The wavy lines $\approx$ instead of the equal sign = mean "almost equal to". Since we no longer have the information for the individual enterprises, but only pooled data are available, we can only approximate the arithmetic mean.

3.3   The Median

The median is also an average value, just defined a little differently. To find the median, we ask what value divides our observations into the 50% smaller observations and the 50%

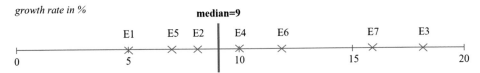

Fig. 3.1 The median for the variable growth

larger observations. To do this, we first need to sort the values in ascending order. For our six companies, then we have the following situation:

$$x_1 = 5$$
$$x_5 = 7$$
$$x_2 = 8$$
$$x_4 = 10$$
$$x_6 = 12$$
$$x_3 = 18$$

The median in the example is then the value that divides our six companies by the growth rate into the 50% companies with the lower growth rate and the 50% companies with the larger growth rate. In this case, it is between 8% and 10%, so the median is simply 9%. We can plot this well on a line (see Fig. 3.1).

Three companies, 1, 5 and 2, have a growth rate smaller than 9%, while companies 4, 6 and 3 have a growth rate larger than 9%.

If we have an uneven number of observations instead of an even number, then the median is exactly the value in the middle. If we add the seventh observation to our example from above and sort them in ascending order, we get the following:

$$x_1 = 5$$
$$x_5 = 7$$
$$x_2 = 8$$
$$x_4 = 10$$
$$x_6 = 12$$
$$x_7 = 16$$
$$x_3 = 18$$

With seven observations we have an uneven number. If we count from the top, then the fourth figure with its value 10 is in the middle, i.e. about 50% of the firms have a growth rate smaller than 10% and about 50% of the firms have a growth rate larger than 10%. We cannot divide the firm in the middle, so there is a small inaccuracy, but we can neglect this.

The median can be used for metric data, as above, as well as for ordinal data. For example, we might have an interest in the founders' self-assessment with regard to their industry professional experience. We collected the data on an ordinal scale from 1 (very experienced) to 5 (not experienced). The evaluation of our first six companies yields a median of 4. We can thus make the statement that the proficiency little experienced divides our company founders in the middle.

3.4   The Mode

The mode, or the most frequent value, is the third average value that we are usually interested in when analyzing data. Basically, the mode can be calculated for metric data, ordinal data, and nominal data. It is most interesting when values occur more frequent. When we look at our data, this is especially the case for ordinal and nominal data. For ordinal data, the mode can complement the median; for nominal data, the mode is the only average value available. In the growth data set, the variables industry, motive, and gender are nominal and the self-assessment variable is ordinal. If we determine the mode for the first six enterprises, we find that the value 1 occurs most frequently in the variable industry. Four newly founded firms are service firms and two are industrial firms. The most frequently mentioned motive for founding is 2. Most founders want to realize an idea. In the case of gender, 0 and 1 each occur three times, i.e. 3 companies were founded by a woman and 3 by a man. For the variable self-assessment, all three values occur equally. The last two examples show that we can also have more than one mode.

With the help of the average values we have already found out a lot.

3.5   The Geometric Mean and Growth Rates

At last, let's turn our attention to the geometric mean. It is used in particular to calculate average growth rates over time. There are plenty of examples of this. We want know to about the average growth rate of the population over the last couple of years, we are interested in the average interest rate of a financial investment, the average profit growth of a company, or average inflation rate over the last years. To find an answer to these things, the geometric mean is needed:

$$\bar{x}_g = \sqrt[n]{x_1 \times x_2 \times \ldots \times x_n}$$

$\bar{x}_g$ is the geometric mean or average growth factor,
x_i are the relative changes from one year to the next, or the growth factors, and
n is the number of growth factors.

To calculate average growth rates, we first need the growth factors, that is, the relative changes from one year to the next. There are two ways to calculate these changes depending on the numbers available. If we have absolute numbers, then we can simply

Table 3.2 Turnover of the first enterprise in the data set from 2007 to 2012

Year	Turnover in CHF	$GR_{t,\%} = \frac{Y_t - Y_{t-1}}{Y_{t-1}} \times 100\%$	$GF_t = 1 + \frac{GR_{t,\%}}{100}$	$GF_t = \frac{Y_t}{Y_{t-1}}$
2007	100'000			
2008	115'000	15%	1.15	1.15
2009	126'500	10%	1.10	1.10
2010	123'970	−2%	0.98	0.98
2011	130'169	5%	1.05	1.05
2012	127,566	−2%	0.98	0.98

divide the absolute numbers by the following formula:

$$GF_t = \frac{Y_t}{Y_{t-1}}$$

GF_t is the growth factor at time t (t for "time" or "point in time"),
Y_t is the value at time t (t can refer to years, months, days, etc.), and
Y_{t-1} is the value at time t − 1.

If only the growth rates are available, then we can use the following formula to calculate the growth factors:

$$GF_t = 1 + \frac{GR_{t,\%}}{100}$$

GF_t is of course again the growth factor at time t and
$GR_{t\%,}$ is the growth rate in percent from time point t − 1 to t.

Here it is good to discuss the formula for the growth rate as well. This will be needed again and again. The formula is:

$$GR_{t,\%} = \frac{Y_t - Y_{t-1}}{Y_{t-1}} \times 100\%$$

It is best to look at all of this in an example. In our data set, we have the average firm growth from 2007 to 2012. These growth rate were calculated the following way. Let's assume we know about the sales figures from 2007 to 2012 (Table 3.2). Using the above formulas (with minor rounding), we receive the annual growth rates and growth factors.

The growth factors are, of course, identical whether they are calculated using the first or the second formula. Now, using the formula for the geometric mean, we come to the average growth factor. This is:

$$\bar{x}_g = \sqrt[5]{1.15 \times 1.10 \times 0.98 \times 1.05 \times 0.98} = 1.05$$

If we subtract 1 from the average growth factor and multiply this by 100%, we receive the average growth rate over the years.

$$\emptyset GR = \left(\overline{x}_g - 1\right) \times 100\% = (1.05 - 1) \times 100\% = 5\%$$

$\overline{x}_g$ is, as mentioned above, the average growth factor and
$\emptyset GR$ is the average growth rate.

Hence, the first observed enterprise grew at an average rate of 5% between 2007 and 2012. Now we understand how the average growth rates of the young enterprises were calculated in our data set data_growth.xlsx.

Better yet, we can use the average growth factor for forecasting purposes, so if we want to know what the company's revenue will be in 2015, all we have to do is to take the last known number and multiply it by the average growth factor for each subsequent year.

$$Turnover_{2015} = Turnover_{2012} \times 1.05 \times 1.05 \times 1.05 = Turnover_{2012} \times 1.05^3 = 147'674 \; CHF$$

But of course this forecast only works well if the average growth factor does not change over time. When we make a forecast, we have to make an assumption. The assumption in our case is that the average growth rate of the past will continue into the future. If this assumption is good, then our forecast is good as well. If the assumption is not so good, then our forecast is not good either.

3.6 What Average Value Should We Use and What else Do We Need to Know?

First, the scale of the variable for which we want to calculate the average value is crucial. If the scale is nominal, only the mode is available. If the scale is ordinal, we can calculate the mode and the median. If the scale is metric, we can calculate the mode and the median as well as the arithmetic mean. If we want to calculate growth rates, we need the geometric mean.

> **Freak Knowledge** Strictly speaking, to calculate the geometric mean we need the presence of a ratio scale. The scale level of metric data can be further divided into an interval scale and a ratio scale. The interval scale requires clearly defined distances between two points. In the ratio scale, an absolute zero point must also be defined. This is not the case with the calendar year, for example, the zero point is the year zero, the birth of Christ. But we could also choose the year zero differently, e.g. death

(continued)

of Ramses II. In the case of turnover, on the other hand, there is an absolute zero point, no turnover. Only when an absolute zero exists two values of a variable can be meaningfully put into a ratio, and only then the geometric mean can be calculated.

In addition, it is important to note that the arithmetic mean is very sensitive to outliers or extreme values, while the median and mode are robust. We can illustrate this again with our cheeseburger example. Suppose we observe the weight of seven cheeseburgers with the following result in grams:

$$x_1 = 251$$
$$x_2 = 245$$
$$x_3 = 252$$
$$x_4 = 248$$
$$x_5 = 255$$
$$x_6 = 249$$
$$x_7 = 47$$

We see that the seventh cheeseburger observed is significantly different from the other cheeseburgers. To illustrate the influence of this extreme value on the arithmetic mean and the median, lets calculate both with and without the extreme value:

$$\bar{x}_{\text{with outlier}} = \frac{\sum x_i}{n} = \frac{251+245+252+248+255+249+47}{7} = 221$$
$$\bar{x}_{\text{without outlier}} = \frac{\sum x_i}{n} = \frac{251+245+252+248+255+249}{6} = 250$$

Before we can calculate the median, we must first sort in ascending order

$$x_7 = 47$$
$$x_2 = 245$$
$$x_4 = 248$$
$$x_6 = 249$$
$$x_1 = 251$$
$$x_3 = 252$$
$$x_5 = 255$$

and then find the value that divides the values into the 50% smaller and 50% larger values. With outliers taken in, the median is 249, without outlier it is 250. The example clearly shows that an extreme value has a large influence on the arithmetic mean, but only a very small influence on the median.

This fact can be used to manipulate. Imagine a lobbyist for a business association is lobbying for higher subsidies. In doing so, he knows that very few companies in his association report very high profits, while the large mass tend to make small or no profits at all. Will he choose the arithmetic mean or the median? He will choose the median because the arithmetic mean is sensitive to the extreme values and is higher than the median. Thus, the arithmetic mean does not meet his goal of getting higher subsidies. If the lobbyist is clever, he will even conceal the fact that he worked with the median, he will only talk about an average. In doing so, he is formally not making a mistake, since all, the mode and the median and the arithmetic mean are called average values. He is just not telling us which average he used. A serious person, on the other hand, will present both averages and draws our attention to the outliers.

3.7   Calculating Averages with Excel

To calculate the average values with Excel, we go to the Formulas tab, click on Insert Function and use the AVERAGE, MEDIAN, MODE.SNGL and GEOMEAN functions.

3.7.1   Calculating the Arithmetic Mean with Excel

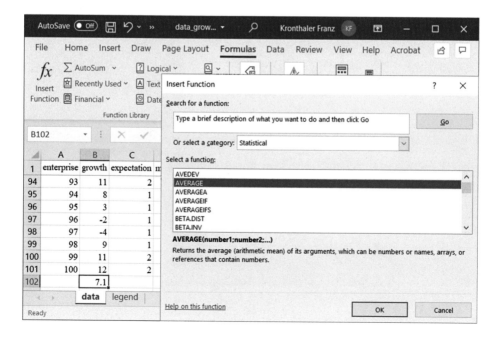

3.7.2   Calculating the Median with Excel

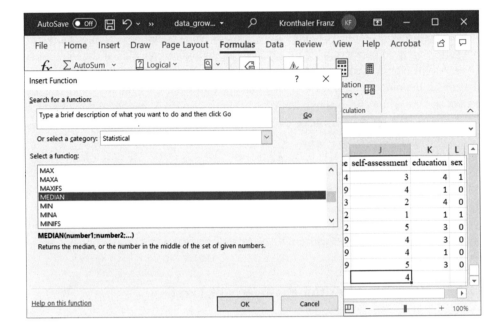

3.7.3   Calculating the Mode with Excel

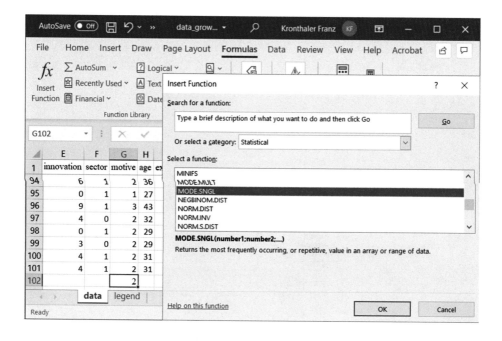

3.7.4   Calculating the Geometric Mean with Excel

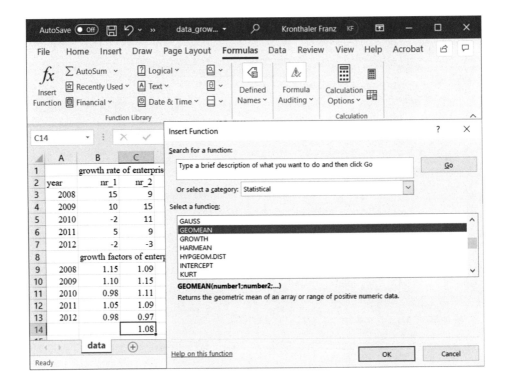

3.8   Checkpoints

- *The mode is used for nominal variables.*
- *For ordinal variables, we can calculate both the mode and the median.*
- *For metric variables, we have the mode, the median, and the arithmetic mean.*
- *The arithmetic mean is often simply called mean.*
- *The arithmetic mean is sensitive to outliers, while the mode and median are not sensitive to extreme values.*
- *The geometric mean is used to calculate average growth rates over time.*
- *Serious people specify which average value they used.*

3.9 Applications

3.1 Which average value should be applied at which scale?

3.2 Calculate by hand for the first six companies of the data set data_growth.xlsx the arithmetic mean for the variables innovation and marketing and interpret the results.

3.3 Calculate by hand for the first six companies of the data set data_growth.xlsx the median for the variables self-assessment and education and interpret the results.

3.4 Calculate by hand for the first six companies of the data set data_growth.xlsx the mode for the variables gender and expectation and interpret the results.

3.5 Is it possible to calculate the median meaningfully for the variable motive?

3.6 For the data set data_growth.xlsx, calculate the average values for all variables using Excel, when meaningful.

3.7 Our third company in the data set grew at 16% in the first year, 11% in the second year, 28% in the third year, 13% in the fourth year, and 23% in the fifth year. What is the average growth rate?

3.8 The turnover of our fourth company is said to have been 120'000 Fr in 2007, 136'800 Fr in 2008, 151'848 Fr in 2009, 176'144 Fr in 2010, 202'566 Fr in 2011 and 193'451 Fr in 2012. Calculate the average growth rate over the years and make a forecast for 2016.

3.9 We are interested in population growth in India and want to compare it with population growth in China. We do some research on the internet and find data for this at the World Bank (https://data.worldbank.org/indicator):

Population at mid-year		
Year	India	China
2013	1'280'846'129	1'357'380'000
2014	1'295'604'184	1'364'270'000
2015	1'310'152'403	1'371'220'000
2016	1'324'509'589	1'378'665'000
2017	1'338'658'835	1'386'395'000
2018	1'352'617'328	1'392'730'000

Source: World Bank 2020, data downloaded from https://data.worldbank.org/ on 11.04.2020.

Calculate the average growth rate for both countries and make a forecast for 2025.

3.10 Which average value is sensitive to outliers, which two are not, and why?

Variation: The Deviation from Average Behavior

4

4.1 Variation: The Other Side of Average Behavior

Average values are only one side of the coin. They are only as good as they represent the persons or objects observed. The other side of the coin is the deviation of the individual observed values from the average. If the individual values are close to the average, then the average value represents the persons or objects well. If the values are further away, then the average value is not so meaningful.

Suppose we observe the salaries of the employees of two small companies, each company with eight employees:

	Employees							
	1	2	3	4	5	6	7	8
Firm 1	6'000	6'000	6'000	7'000	7'000	8'000	8'000	8'000
Firm 2	4'000	4'000	4'000	4'000	4'000	8'000	8'000	20'000

Employee 1 of firm 1 earns CHF 6'000 per month, employee 2 of firm 1 earns the same, and so on. If we calculate the average salary of employees for both firms using the arithmetic mean, we find that the average salary in both firms is CHF 7'000:

$$\bar{x}_1 = \frac{\sum x_i}{n} = \frac{6'000 + 6'000 + 6'000 + 7'000 + 7'000 + 8'000 + 8'000 + 8'000}{8} = 7'000$$

$$\bar{x}_2 = \frac{\sum x_i}{n} = \frac{4'000 + 4'000 + 4'000 + 4'000 + 4'000 + 8'000 + 8'000 + 20'000}{8} = 7'000$$

© Springer-Verlag GmbH Germany, part of Springer Nature 2023
F. Kronthaler, *Statistics Applied With Excel*,
https://doi.org/10.1007/978-3-662-64319-8_4

If we compare the average salary with the individual salaries of the employees, we immediately see that the mean value of 7'000 francs reflects the salaries of the first company well, that is, the average is meaningful here. For the second company, however, the salaries deviate a lot from the mean. The mean value is not so informative about the salaries of the individual employees.

Variation or the deviation of individual observations from the average, is a central concept in statistics on which almost all methods of statistics are based on. It's worth exploring the concept of variation more in depth. Once we understand it, many things become easier. To analyze variation, the range, the standard deviation, the variance, the coefficient of variation, the interquartile range as well as the boxplot are good tools.

4.2   The Range

The simplest measure of variation is the range. The range ra is the distance between the largest and smallest value of a variable:

$$ra = x_{max} - x_{min}$$

It can be calculated for ordinal and metric data and gives a quick overview of the area in which the observed values are. In our example growth of young enterprises, the range of the growth rate for the first six firms is 13 percentage points (see Table 1.5).

$$ra = x_{max} - x_{min} = 18 - 5 = 13$$

Whether 13 percentage points difference in the growth rate is a lot or not so much is a matter of interpretation and we'll put that aside for now. But we now know that the values are within this range. The range is perhaps more impressive for observed salaries. Suppose we observe the monthly salaries of 100 people. The observed minimum salary is CHF 3'500, but it just happens that Bill Gates is among them. Completely fictitious he earns CHF 50 million. Then the range is an impressive high number:

$$ra = x_{max} - x_{min} = 50'000'000 - 3'500 = 49'996'500$$

The range gives a quick overview over the area in which the values are. Furthermore, it gives hints to unusual numbers, to outliers (Bill Gates). At the same time, it is very sensitive to outliers, since only the smallest and largest values are considered. A measure that includes all values when calculating the deviation is the standard deviation.

4.3 The Standard Deviation

The standard deviation can be interpreted as the average deviation of individual observations from the mean. The calculation assumes metric data. We first calculate the deviations of each value from the mean, square these deviations (so that negative and positive deviations do not add up to zero), divide by the number of observations, and then take the square root:

$$s = \sqrt{\frac{\sum (x_i - \bar{x})^2}{n - 1}}$$

For our example growth of firms, we have the following growth rates:

$$x_1 = 5$$
$$x_2 = 8$$
$$x_3 = 18$$
$$x_4 = 10$$
$$x_5 = 7$$
$$x_6 = 12$$

The arithmetic mean is:

$$\bar{x} = 10$$

The standard deviation can be calculated then as follows:

$$s = \sqrt{\frac{\sum (x_i - \bar{x})^2}{n - 1}} = \sqrt{\frac{(5-10)^2 + (8-10)^2 + (18-10)^2 + (10-10)^2 + (7-10)^2 + (12-10)^2}{6-1}}$$

$$\sqrt{\frac{(-5)^2 + (-2)^2 + 8^2 + 0^2 + (-3)^2 + 2^2}{5}} = \sqrt{\frac{25 + 4 + 64 + 0 + 9 + 4}{5}}$$

$$= \sqrt{\frac{106}{5}} = \sqrt{21.2} = 4.6$$

The average growth rate is 10%, with an average deviation of the observed values of 4.6 percentage points. The deviation is thus 46% of the value of the mean. Is this a lot or a little? The interpretation depends on our expertise regarding the growth rates of young enterprises. A banker who deals with the financing of young enterprises might, for example, come to the conclusion that growth rates do fluctuate considerably, which would then have an impact on his financing decisions. He might also wonder about the reasons for the fluctuations.

Freak Knowledge

As we did with the arithmetic mean $\bar{x}$ we calculated the standard deviation for the sample and denoted it by s. When we calculate the standard deviation for the population, i.e. all objects or persons in question, we do not divide by $n-1$ but only by n. In addition, the standard deviation is then denoted by the Greek letter σ and the number of observations with N. The question may arise why we divide by $n-1$ for the sample. In particular, if the sample is small, the standard deviation will be underestimated. If we divide by n − 1 instead of by n, we correct for this fact.

If we already have the data prepared in a classified way, as we discussed with the calculation of the arithmetic mean too, then we need to adjust the formula slightly. For this, we consider again Table 3.1, which is shown again below.

Class	Growth rate from over ... to ... [in percent]	Number of enterprises
1	−10 to −5	3
2	−5 to 0	8
3	0 to 5	27
4	5 to 10	34
5	10 to 15	23
6	15 to 20	4
7	20 to 25	1

The formula is then:

$$s \approx \sqrt{\frac{\sum (\bar{x}_j - \bar{x})^2 \times n_j}{n-1}}$$

$\bar{x}_j$ are the mean values of the respective classes,
n_j is the number of companies in the respective class, and
n the total number of observed firms.

In class one with three firms, the mean of the growth rate is −7.5, in class two with eight firms −2.5, and so on. We have already discussed the mean value in Chap. 3, which is

$\bar{x} \approx 6.6\%$. We can thus calculate the standard deviation as follows:

$$s \approx \sqrt{\frac{\sum (\bar{x}_j - \bar{x})^2 \times n_j}{n - 1}}$$

$$\approx \sqrt{\frac{(-14.1)^2 \times 3 + (-9.1)^2 \times 8 + (-4.1)^2 \times 27 + 0.9^2 \times 34 + 5.9^2 \times 23 + 10.9^2 \times 4 + 15.9^2 \times 1}{100 - 1}}$$

$$\approx \sqrt{\frac{596.43 + 662.48 + 453.87 + 27.54 + 800.63 + 475.24 + 252.81}{99}} \approx \sqrt{\frac{3269}{99}} \approx 5.8$$

The average growth rate of firms over the past 5 years is around 6.6% with a standard deviation of 5.8 percentage points.

4.4   The Variance

One more important thing we need to know. If we don't take the root in the standard deviation formula, we get the variance instead of the standard deviation. Likewise, of course, we can square the standard deviation:

$$\text{var} = \frac{\sum (x_i - \bar{x})^2}{n - 1}$$

The variance, the squared standard deviation, is more difficult to interpret in terms of content than the standard deviation. We do not only square the values, but the units as well. In the example calculated above with unclassified values, the variance (4.6 percentage points)^2 is 21.16 percentage points^2. However, percentage points^2 cannot be interpreted any more, like to give an other example CHF^2. Nevertheless, the variance is important. As stated earlier, deviation is a central concept used in statistical methods. When applying these methods, we usually don't bother about taking the root, but use the variance as a measure of variation instead.

Whenever you hear the term variance, you should think of deviation from average behavior. In one of the next chapters, the chapter on correlation, we will hear the term covariance. Again, deviation is the matter of fact, with "co" used to refer to the joint deviation of two variables.

4.5   The Coefficient of Variation

The coefficient of variation is used when we compare the variation of two variables, that is, when we wonder in which variable people or objects have the larger deviation or are less consistent. It measures the variation relatively, as a percentage of the mean:

$$cv = \frac{s}{\overline{x}} \times 100$$

cv is the coefficient of variation,
s is the standard deviation, and
$\overline{x}$ is the mean.

Let's take the example of the growth rates of our six companies again. We calculated the following mean and standard deviation:

$$\overline{x} = 10 \quad \text{and} \quad s = 4.6$$

In the same way, we can calculate the mean and standard deviation for the variable age. If we do this, we get:

$$\overline{x} = 38 \quad \text{and} \quad s = 8.2$$

Do the values for the variable age deviate more from the mean value than the values for the variable growth? At least the standard deviation is larger, but furthermore the mean too. If we calculate the coefficient of variation of both variables, we get the following values:

$$cv_{\text{growth rate}} = \frac{s}{\overline{x}} \times 100 = \frac{4.6}{10} \times 100 = 46.0$$

$$cv_{\text{age}} = \frac{s}{\overline{x}} \times 100 = \frac{8.2}{38} \times 100 = 21.6$$

Comparing the values, it can be seen that the standard deviation for the variable age is 21.6% of the mean, whereas for the variable growth rate it is 46%. Using the coefficient of variation, we see that for the variable growth the deviation of the individual observations from their mean is larger than for the variable age. The mean of the variable age thus better represents the individual persons or in other words the observed companies are more consistent in age of the founder than in their growth rate.

4.6   The Interquartile Range

Another important measure of variation is the interquartile range. It is used for both metric data and ordinal data and informs about the area in which the middle 50% of the values are located.

For simplicity reasons, we expand our example from 6 to 8 enterprises. The growth rates of the first eight firms in our data set are as follows:

$$x_1 = 5$$
$$x_2 = 8$$
$$x_3 = 18$$
$$x_4 = 10$$
$$x_5 = 7$$
$$x_6 = 12$$
$$x_7 = 16$$
$$x_8 = 2$$

To determine the interquartile range, we must first sort the values in ascending order, just as we did with the median:

$$x_8 = 2$$
$$x_1 = 5$$
$$x_5 = 7$$
$$x_2 = 8$$
$$x_4 = 10$$
$$x_6 = 12$$
$$x_7 = 16$$
$$x_3 = 18$$

The interquartile range is then the value that reflects in which area the four middle companies are located. In our case, the area is from 6% to 14%. These two values are also called first quartile and third quartile. They are the middle values between the two values that are closest to the respective quartile. Thus, the interquartile range is 8 percentage points. The median is also called the 2nd quartile. All this can also be well represented on a line (Fig. 4.1).

If we look at the quartiles again, we notice that there are in this case two values between the quartiles, or 25% of the values. Accordingly, between the first quartile and the third quartile there are 50% of the values.

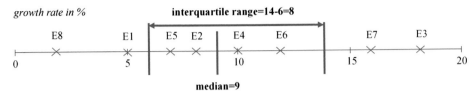

Fig. 4.1 The interquartile range for the variable growth

Freak Knowledge
In the above example, we calculated the first quartile as $(5 + 7)/2 = 6$ and the
third quartile as $(12 + 16)/2 = 14$. For the calculation, we said that these values are
obtained from the middle of the two values closest to the quartile. However, when the
quartiles are calculated using Excel, it usually results in slightly different numbers,
which can be confusing. Why is this? There are different methods of calculating
the quartiles. For example, Excel weights the neighboring values according to
the following formulas: $Quartile_{(0.25)} = X_{Index} + Weight * (X_{Index+1} - X_{Index})$ or
$Quartile_{(0.75)} = X_{Index} + Weight * (X_{Index+1} - X_{Index})$, where the weight is the rank
minus the index (Weight: $=$ Rank index). The rank per quartile is defined either as
rank: $=0.25*(n-1)+1$ or as rank: $=0.75*(n-1)+1$ of the sorted values and the
index is the whole number of the respective rank, i.e. we cut off the number of
the index after the decimal point. If we calculate the quartiles for our example
accordingly, then we get the value 6.5 for the first quartile and the value 13 for
the third quartile, just like in Excel. The interested reader can try to recalculate the
values. For us in the application, however, it is only important that 25% of the values
are smaller than the first quartile and 75% of the values are smaller than the third
quartile, regardless of how the values are calculated exactly.

We have learned about measures of variation for metric and ordinal variables, but not for
nominal variables. We do not need measures of variation for nominal variables. We can
illustrate this with the example of gender. Men and women like to stay close to each other,
but they do not scatter around the opposite sex. Technically speaking, we have zeros (men)
and ones (women), but no intermediate values.

4.7 The Boxplot

The boxplot combines the information on the median, the first and third quartiles as well as
the minimum and maximum values. The boxplot is therefore a popular tool to graphically
analyze the variation of ordinal and metric variables. The following figure shows the
boxplot for our first eight companies (Fig. 4.2).

Fig. 4.2 Boxplot variable
growth

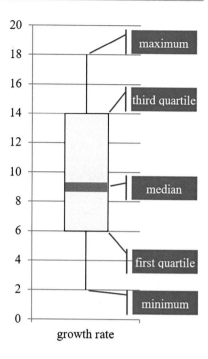

Among the considered companies, the lowest growth rate achieved by a company is
2%, the highest is 18%, the first quartile is 6%, the third quartile is 14% and the median
is 9%. All this information is shown in the boxplot. The height of the box is defined as
the difference between the third quartile and the first quartile. Within this range the middle
50% of the firms are located. The ends of the lines mark the minimum and maximum
growth rates. Between the first quartile and the end of the line are thus the 25% of the
companies with the lowest growth rates. Between the third quartile and the end of the
line are the 25% of companies with the highest growth rates. The median divides our
companies exactly into the 50% with the higher growth rates and the 50% with the lower
growth rates.

Freak Knowledge

The location of the median also provides information about whether a variable is
symmetric or skewed. If the median divides the box exactly in the middle, then we
are talking about a symmetric variable. If the median is in the upper half of the box,
the variable is left skewed. Similarly, if the median is in the lower half of the box,
the variable is right skewed.

(continued)

In addition, the boxplot often marks unusual values with an asterisk or a dot. The idea is that all values outside a certain range are unusual observations and should not be integrated in the range from the minimum to the maximum. One possibility that is often used is to limit the lines (also called whiskers) to 1.5 times the length of the box.

4.8   Calculating Variation Measures with Excel

To calculate the measures of deviation with Excel, we have the following functions available under the Formulas tab: MIN, MAX, STDEV.S, VAR.S. and QUARTILE.INC.

4.8.1   Calculating the Range (MIN, MAX) with Excel

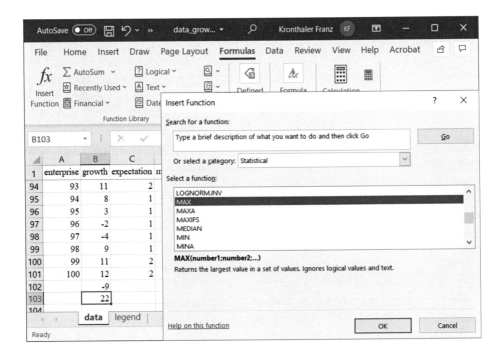

The range is the difference between the two values.

4.8.2   Calculating the Standard Deviation with Excel

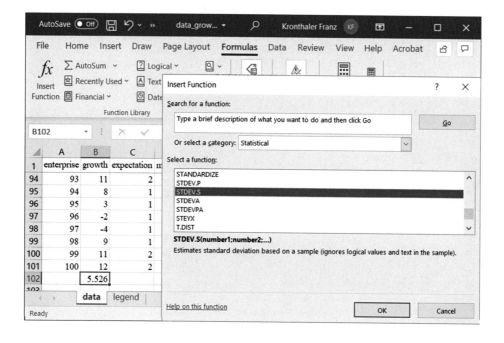

4.8.3   Calculating the Variance with Excel

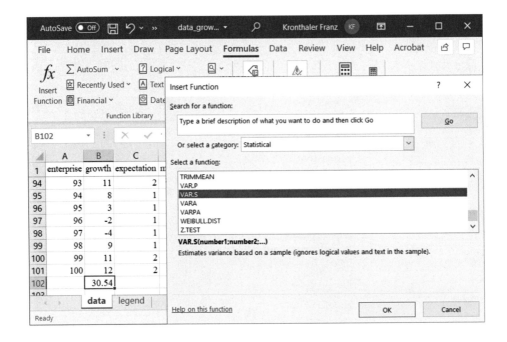

4.8.4 Calculating of the Interquartile Range (First Quartile and Third Quartile) with Excel

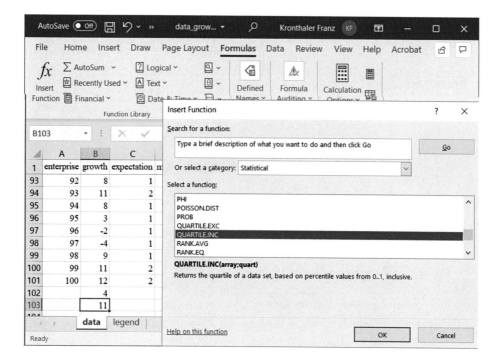

To calculate the first quartile enter 1, to get the third quartile enter 3. The difference of both values is the interquartile range.

4.9 Creating the Boxplot with Excel

To create a boxplot, the Box and Whisker diagram is available in Excel 2019. To draw one, it is best to copy the data we need from the original data set. Afterwards, we sort them by the groups for which we want to draw the boxplots. In the Insert tab, we then select the Box and Whisker diagram from all of the charts. The following figure summarizes the procedure. For special formatting, please refer to the Excel help function.

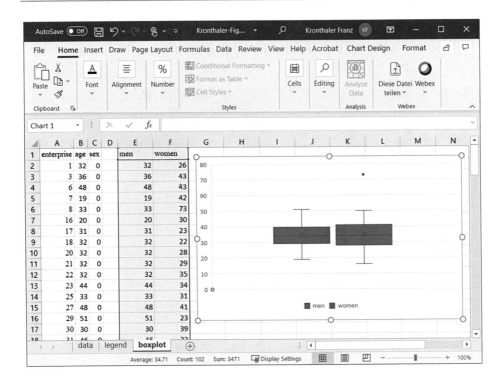

For this chart, the original data set has been slightly modified. The age of one female founder was increased to 73, so that an outlier is visible at the top of the chart.

4.10   Checkpoints

- Measures of variation complement our information on the average; the smaller the variation, the more meaningful the average value is.
- The range is the difference between maximum and minimum of a variable, it is very sensitive to outliers.
- The standard deviation can be interpreted as the average deviation of all values from the mean. The variance is the squared standard deviation. For metric variables, both values can be calculated in a meaningful way.
- The interquartile range gives information about the area in which the middle 50% of the values are located. It can be used with both metric and ordinal data.
- The coefficient of variation is applied to analyze in which variable people or objects are more consistent, it is applied with metric data.
- The boxplot is a popular graphical tool to analyze variation.

4.11   Applications

4.1 Which measure of variation is useful at which scale?

4.2 For nominal data, is there a measure of variation, why or why not?

4.3 For the first eight companies in our data set data_growth.xlsx, calculate by hand the range, the standard deviation, the variance, the interquartile range, and the coefficient of variation for the variable age and interpret the result.

4.4 For the first eight companies in our data set data_growth.xlsx, calculate by hand the range, the standard deviation, the variance, the interquartile range, and the coefficient of variation for the variable marketing and interpret the result.

4.5 For the first eight companies in our data set data_growth.xlsx, calculate by hand the range, the standard deviation, the variance, the interquartile range, and the coefficient of variation for the variable innovation and interpret the result.

4.6 In which variable are the enterprises less consistent, marketing or innovation (applications 4.4 and 4.5)? What measure of variation do you use to answer the question?

4.7 We are interested in the volatility of two shares, that is, how much the share prices fluctuate over time. We look for a financial platform on the Internet that provides the data and look at the daily closing prices for one week (e.g. https://www.finanzen.ch). For the first share (Novartis) we get the following values: 16.03 CHF 70.19, 17.03 CHF 73.49, 18.03 CHF 72.02, 19.03 CHF 75.70, 20.03 CHF 74.00. For the second share (UBS) we get: 16.03 CHF 7.46, 17.03 CHF 7.76, 18.03 CHF 7.77, 19.03 CHF 8.73, 20.03 CHF 8.26. Calculate the range, the standard deviation, the variance and the coefficient of variation and decide which share had a higher volatility in the given period.

4.8 We have the following values for the variable growth from our data set: minimum −9, maximum 22, first quartile 4, median 8, third quartile 11. Draw the boxplot by hand. What is the range of the middle 50% of the observations?

4.9 Using Excel, draw the boxplot for the variable growth of the data set data_growth.xlsx.

4.10 We have the following values for the variable growth, grouped by motive for starting a business, from our data set: unemployment: minimum −3, maximum 22, first quartile 2, median 7, third quartile 10; implement idea: minimum −7, maximum 20, first quartile 4.25, median 8, third quartile 11; higher income: minimum −9, maximum 13, first quartile 3, median 7, third quartile 11. Draw the boxplots and compare the variation.

4.11 Use Excel, draw the boxplots for the variable growth by motive for starting an enterprise.

4.12 For our data set data_growth.xlsx, calculate the range, the interquartile range, the standard deviation, and the variance for all variables using Excel (if possible). Also, draw the associated boxplots.

Charts: The Possibility to Display Data Visually

5

5.1 Charts: Why Do We Need Them?

We have already learned about some statistical tools, the arithmetic mean, the median, the mode, the geometric mean, the range, the standard deviation, the variance and the interquartile range. With these, we have calculated how people or objects behave on average and how they deviate from average behavior. Hence, we know already a lot about our persons or objects. Experience shows, however, that numbers are sometimes difficult to communicate. People are quickly bored if they hear and see only numbers. This is an important reason why we need statistical charts. Charts helps to display numbers and create lasting impressions.

But graphs also help to see patterns in the data, or give hints about the behavior of people or objects, such as how many people like to eat cheeseburgers a lot, or how many companies have growth rates between 10% and 15%, how many of the company founders are very experienced in their industry, etc. To discuss such questions, we can use the frequency table and frequency charts.

5.2 The Frequency Table

The frequency table is the starting point, it shows how often values occur in absolute and relative terms. We distinguish frequency tables according to the scale of the data. With nominal and ordinal data, usually only a few specific values occur and we can enter the values directly into the table. With metric data, usually we have a lot of different values, i.e. first we have to classify them. Let's start with the frequency table for nominal and ordinal data.

© Springer-Verlag GmbH Germany, part of Springer Nature 2023
F. Kronthaler, *Statistics Applied With Excel*,
https://doi.org/10.1007/978-3-662-64319-8_5

Table 5.1 Frequency table for the variable self-assessment

Self-assessment	Frequency, absolute	Frequency, relative	Frequency, cumulative
x_i	n_i	f_i	cf_i
1	6	0.06	0.06
2	17	0.17	0.23
3	21	0.21	0.44
4	30	0.30	0.74
5	26	0.26	1
	$n = 100$		

Table 5.1 displays the frequency table for the (ordinal) variable self-assessment for the 100 enterprises observed. The values range from 1 (very experienced) to 5 (not experienced).

x_i	are in this case the values the variable can take,
n_i	is the number of occurrences of the respective value or the absolute frequency,
$f_i = \frac{n_i}{n}$	is the number of each value to the total number of observations or the relative frequency, and
cf_i	is the cumulated relative frequency.

If we multiply f_i resp. cf_i by 100, we get percentage values. The table is easy to interpret. $x_1 = 1$ occurs six times. Accordingly, 6 out of 100, i.e. 6% of the companies are managed by a founder with a lot of industry experience. $x_2 = 2$ occurs 17 times, i.e. 17% of all companies. If we add up the two values, we arrive at the cumulative share of 0.23 or 23%, i.e. 23% of the companies are led by a founder with a lot of industry experience or by a founder with experience in the industry.

As mentioned already, when we have metric data, we should classify the data first. The reason for this is that metric values usually do not occur more often and forming frequencies without classifying them makes no sense. For the classification we first look at the range and then consider how many classes should we make with what width. Too many classes are not good, nor are too few. We usually work with 5 to 15 classes. For our variable growth, the range is from −9 to 22%, which is 31 percentage points. We quickly see that we can work well with a class width of 5 percentage points, which results in seven classes (Table 5.2).

We can interpret the table quite similarly to Table 5.1. Three companies have a growth rate between −10 and −5%. That is 3% of the companies. Another eight companies, or 8% of the companies, have a growth rate between −5 and 0%, and so on. If we add both values together, we see that a total of 11% of the companies have a growth rate of less than or equal to 0%. The issue that we count from above a value to a specific value (see first column of Table 5.2) has to do with the fact that we want to avoid double counting.

Table 5.2 Frequency table for the variable growth

Growth	Frequency, absolute	Frequency, relative	Frequency, cumulative
from over ... to ...	n_i	f_i	cf_i
-10 to -5	3	0.03	0.03
-5 to 0	8	0.08	0.11
0 to 5	27	0.27	0.38
5 to 10	34	0.34	0.72
10 to 15	23	0.23	0.95
15 to 20	4	0.04	0.99
20 to 25	1	0.01	1
	$n = 100$		

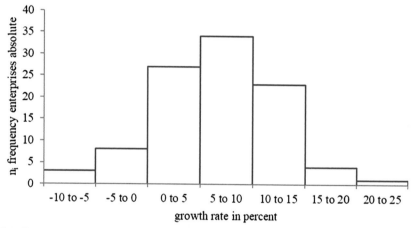

Fig. 5.1 Frequency chart with absolute frequencies for the variable growth

5.3 The Frequency Charts

There are three possibilities to display the information of the frequency table. Either we draw an absolute frequency chart, a relative frequency chart or a histogram.

In the absolute frequency chart (Fig. 5.1), we display the absolute frequencies n_i on the y-axis. On the x-axis we plot the classes for metric data or the possible values for nominal or ordinal data. Hence, 3 companies reflect the height of the first bar, 8 enterprises reflect the height of the second bar, and so on.

In the relative frequency chart (Fig. 5.2), we plot on the y-axis the relative frequencies f_i and on the x-axis the classes for metric data or the possible values for ordinal or nominal data. The height of the bars show that 0.03 or 3% of the enterprises are included in the first class, 0.08 or 8% of the companies in the second class, and so on.

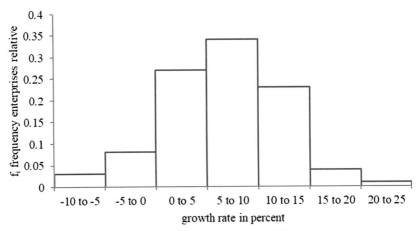

Fig. 5.2 Frequency chart with relative frequencies for the variable growth

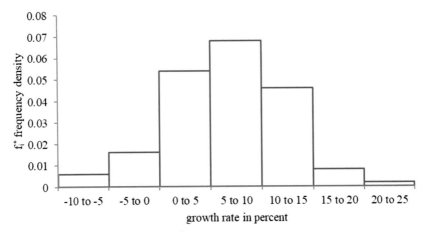

Fig. 5.3 Histogram for the variable growth

In the histogram (Fig. 5.3), we plot on the y-axis the frequency density, where w_i is the class width of the respective class. On the x-axis we display the classes or the possible values.

$$f_i^* = \frac{f_i}{w_i}$$

The difference to the absolute or relative frequency chart is that it is no longer the height of the bar that indicates what percentage of the companies are in the respective class, but the size of the area. We can easily show this by solving the formula of the frequency density for f_i. The relative frequency is the frequency density times the respective class width $f_i = f_i^* \times w_i$.

5.4   Absolute Frequency Chart, Relative Frequency Chart, or Histogram?

Which frequency chart should we choose? When we have look at the three chart, we notice that they are visually identical. This is always the case when all classes have the same width. Here it makes no visual difference which chart we choose. However, the absolute frequency plot and the relative frequency plot are easier to interpret, as the height of the bars reflects the number or proportion of companies.

If we have different class widths then the histogram is preferable. The histogram takes into account the different class widths, by dividing the relative frequency by the class width. If we would not do this, the bar would only get higher because the class is wider. So if we use the absolute or relative frequency chart, different class widths will give us a distorted picture, whereas the histogram will give us the undistorted, correct visual picture. We have different class widths especially when our range is influenced by extreme values. Including all values in such a case, the same class widths would require to create too many classes. If we would like to avoid this, we must either have to exclude values or vary the class width.

> **Freak Knowledge**
> The graphical frequency chart, like the boxplot, is a tool to evaluate whether a variable is symmetric. It is symmetric if we cut out the frequency chart, fold it in the middle and both sides come to rest on each other. Furthermore, we use the frequency chart to assess whether a variable is normally distributed. It is normally distributed when the frequency plot looks like the Gaussian bell curve.

5.5   More Ways to Display Data

The frequency charts are not the only ways to present data. Before we show other easy possibilities, here are a few general hints to help to create meaningful charts.

To produce a good graph, the following rules should be followed:

1. Creating a good chart is not easy, it is an art.
2. Before creating a chart, it should be clear what information should be displayed.
3. As a rule, only one piece of information should be presented in a chart.
4. A good chart is as simple and understandable as possible.

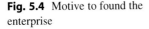 **Fig. 5.4** Motive to found the enterprise

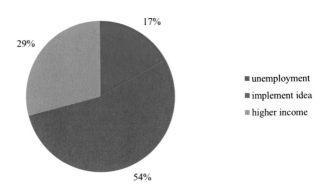

5. A good chart is not misleading.
6. A good chart complements the text and facilitates its interpretation.
7. If more than one chart are used in a text, they should all be consistent in formatting, e.g., size, font size, type style, and so on.

We will discuss the pie chart, the bar chart, the line chart and the map. Of course, this does not exhaust the various ways to create graphs. Anyone interested in displaying data beyond this book should look at the work of Edward R. Tufte (Tufte E. R. (2001), The Visual Display of Quantitative Information, 2ed, Graphics Press, Ceshire, Connecticut). Edward Tufte has been called the "Leonardo da Vinci of data" by the New York Times.

The pie chart is, for example, used when we would like to show how many people or objects have a certain characteristic. Figure 5.4 shows from which motives the enterprises in our data set were founded. The majority of the founders, 54%, wanted to implement an idea, 29% pursue the goal of achieving a higher income and unemployment was the decisive factor for 17% of the enterprise foundations.

What we see here too, is the two-dimensionality of the chart. In science, we avoid unnecessary formatting, as it merely distracts from the facts to display. In consulting, on the other hand, graphs are often drawn in three dimensions to make the information more appealing. However, without a good reason unnecessary formatting should be avoided.

The bar chart is of good use when the goal is to compare groups, for example, when we look at men and women in terms of their motive to found the enterprise. Figure 5.5 shows that male and female founders differ slightly in their motivation to start the enterprise. 18.5% of the founders set up a company out of unemployment, whereas only 14.3% of the female founders did so. On the other hand, 60% of female founders wanted to implement an idea, while this is the case for only 50.8% of male founders. With the motive of achieving a higher income, again more men (30.8%) are represented in percentage share than women (25.7%).

The line chart is used in particular when we want to show a trend. Figure 5.6 shows the turnover development of two enterprises in our data set from 2007 to 2012. Here, we observe how the turnover of company no. 1 has developed compared to company no. 5.

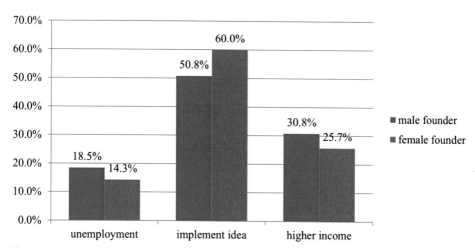

Fig. 5.5 Motive to start the enterprise by sex

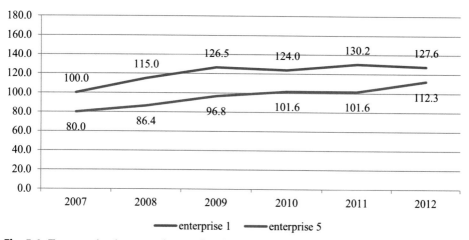

Fig. 5.6 Turnover development of enterprises from 2007 to 2012 (in CHF 1000)

Both tend to show a positive trend. However, it appears that the turnover of the enterprise no. 1 stays roughly the same since 2009, while the turnover of enterprise no. 5 have been increasing throughout the period, with one small exception.

Finally, let's look at the map. Excel offers the possibility to create a map in the newer versions. We use maps when we want to present geographical data. Many facts differ by region and we can show this very memorably with a map chart. The following chart shows the real gross domestic product per capita of European countries (source: Eurostat, 2020). We can see relatively quickly, which regions in Europe belong to the richer regions and which regions belong to the poorer ones (Fig. 5.7).

Fig. 5.7 GDP per capita of European countries in 2019

5.6   Creating the Frequency Table, Frequency Charts and Other Graphs with Excel

To create the frequency table, the FREQUENCY function is available.

Before we apply the function, it is useful to prepare the table. To do so, we set the class borders (Excel requires only the upper values) and label the table.

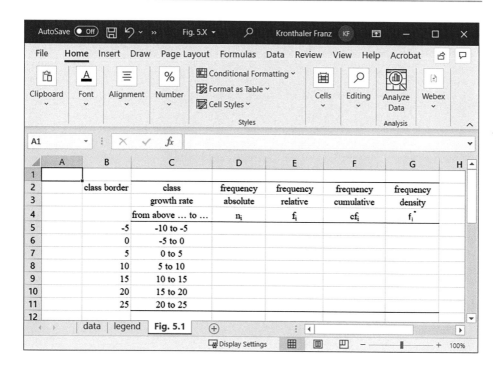

Then we use the Frequency function in the tap Formulas. To do this, we first select all cells for which we want to have the frequency, then click on the Formulas tab, and look by Insert function for the function Frequency.

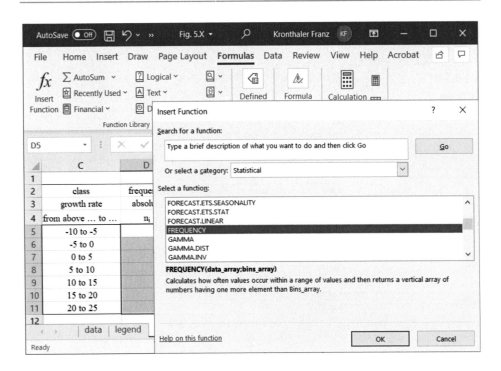

After we have clicked on OK, the following window opens, in which we enter the data range and the upper class borders.

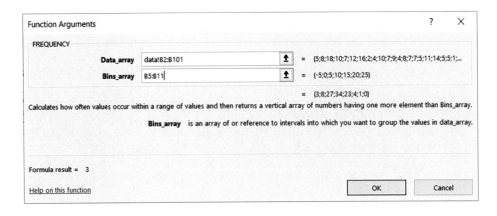

Then we click OK and the following screen appears:

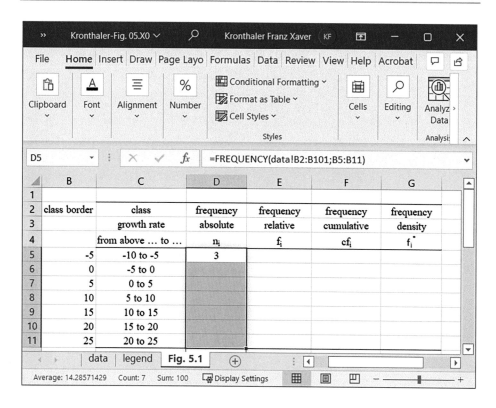

To get the frequency for all cells that are selected, we click after the Frequency formula in the function row.

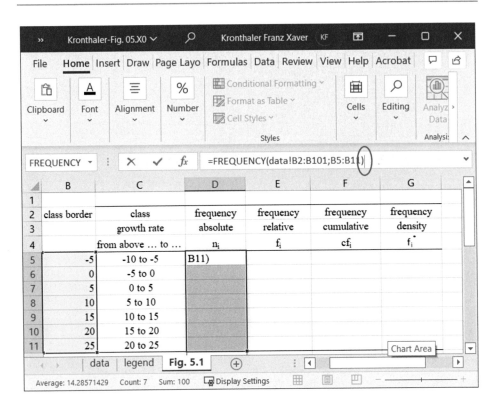

When the cursor flashes behind the formula, we press the control, shift and return keys and we get the frequencies.

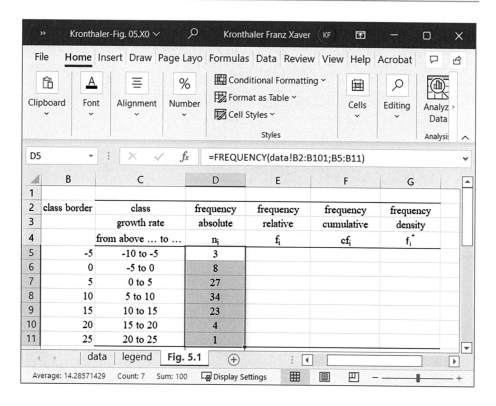

The last thing to do is calculate the remaining fields.

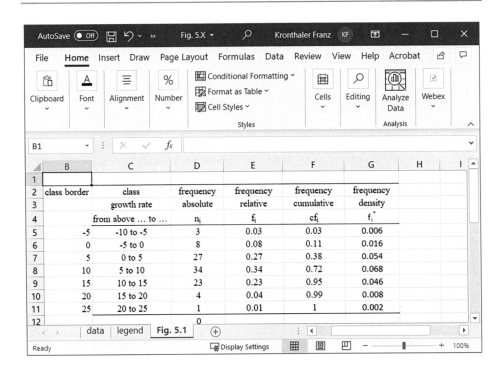

class border	class growth rate from above ... to ...	frequency absolute n_i	frequency relative f_i	frequency cumulative cf_i	frequency density $f_i^{\cdot}$
-5	-10 to -5	3	0.03	0.03	0.006
0	-5 to 0	8	0.08	0.11	0.016
5	0 to 5	27	0.27	0.38	0.054
10	5 to 10	34	0.34	0.72	0.068
15	10 to 15	23	0.23	0.95	0.046
20	15 to 20	4	0.04	0.99	0.008
25	20 to 25	1	0.01	1	0.002
		0			

When the table is ready, we can draw the frequency chart. To do so, we select the necessary data and then click on recommended charts. Here we select the correct chart and click on OK.

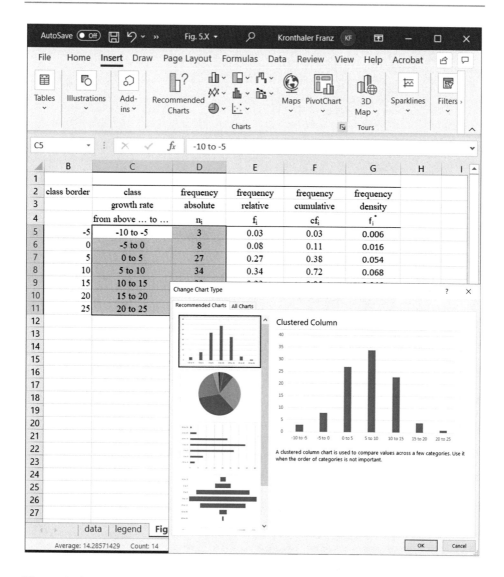

We get the following picture.

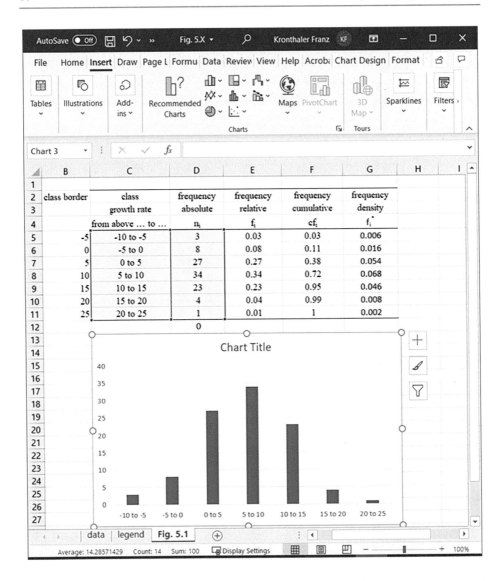

Now we can finish our frequency chart by labeling the axes, reducing the bar spacing, and so on.

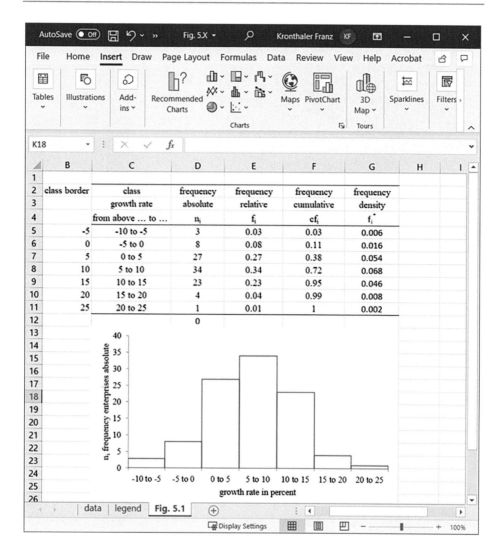

class border	class growth rate from above … to …	frequency absolute n_i	frequency relative f_i	frequency cumulative cf_i	frequency density f_i^*
-5	-10 to -5	3	0.03	0.03	0.006
0	-5 to 0	8	0.08	0.11	0.016
5	0 to 5	27	0.27	0.38	0.054
10	5 to 10	34	0.34	0.72	0.068
15	10 to 15	23	0.23	0.95	0.046
20	15 to 20	4	0.04	0.99	0.008
25	20 to 25	1	0.01	1	0.002
		0			

To create another chart instead of the column chart, we select the appropriate chart under the Insert tab. It is important that the data is prepared in advance, i.e. counted and entered correctly, etc. In the following, this is briefly shown for the statistical charts shown above. I will not go into details here. If needed, now is the time to use Excel's help function.

One more important point: we should not forget to label the graphs sufficiently with titles and axes. A graph whose details are not clear is worthless. Unfortunately, we see such graphs all the time.

To draw a pie chart, we count the data using the function Frequency. Then we select the Pie chart under the tab Insert.

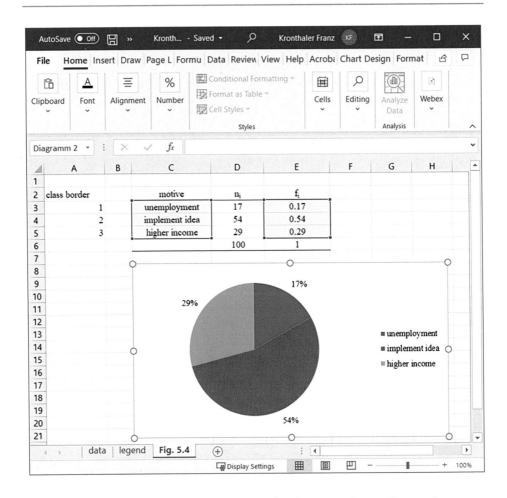

In order to draw a bar chart, the values must first be counted according to the groups. Furthermore, due to comparability, it is always worth considering whether it is better to express the values as percentages, especially if the group or sample sizes are different.

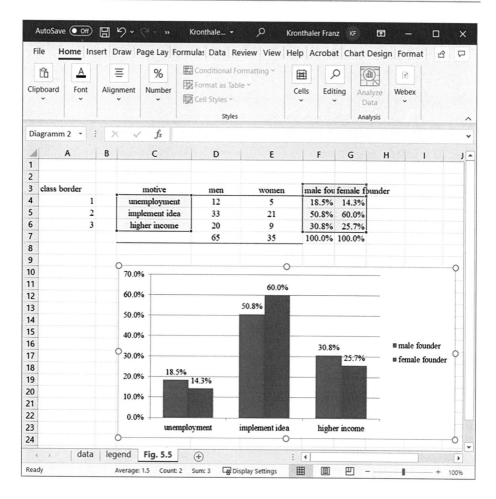

Fig. 5.5

In a line chart, it is best to arrange the trend lines to be drawn next to each other in the columns.

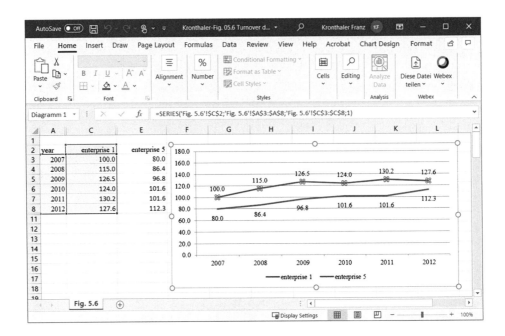

A map chart can be drawn using the newer versions of Excel. To do this, the respective regions must be entered in the first column and the values to be displayed in the column next to it. Then the Maps button can be selected in the tab Insert and the data entered. It is important that a working internet connection is available and that the region names entered meet the respective country standards.

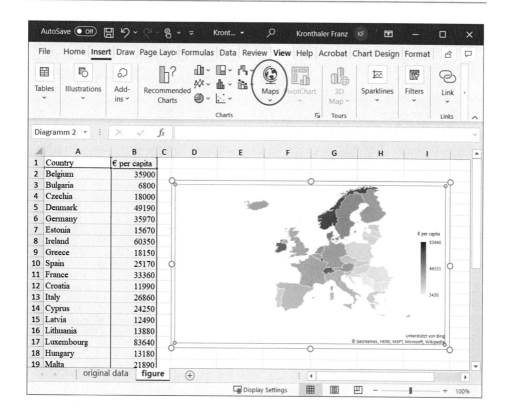

5.7   Checkpoints

- *Graphical presentation of data helps to explain facts.*
- *Graphical presentations of data often give hints about the behavior of people or objects, as well as patterns in the data.*
- *We use frequency charts to display how often certain values occur.*
- *The frequency table shows how often certain values occur in absolute and relative terms.*
- *Graphically, we can represent the frequency distribution absolutely, relatively, or with the histogram.*
- *If we have equal class widths, the absolute or relative frequency plot is preferable; if we have different class widths, the histogram is preferable.*
- *A good chart is as simple as possible and presents information clearly.*

5.8 Applications

5.1 Calculate the average growth rate of the enterprises in Fig. 5.6. To do this, read the numbers from the figure. Compare the calculated growth rates with the respective growth rates from our data set data_growth.xlsx.

5.2 Give two reasons why charts are helpful.

5.3 For what purpose do we need frequency charts?

5.4 For different class widths, is the absolute frequency chart, the relative frequency chart, or the histogram preferable? Why?

5.5 For the variable expectation for the first six companies in our data set data_growth.xlsx, create the frequency table, absolute and relative frequency chart by hand and interpret the result.

5.6 For our data set data_growth.xlsx, create the frequency table and the relative frequency chart for the variables marketing and innovation using Excel and interpret the result.

5.7 Is the marketing variable more symmetric or the innovation variable more symmetric? Why?

5.8 Create the relative frequency chart and the histogram for the variable growth using the following class widths: -10 to -5, -5 to 0, 0 to 5, and 5 to 25. Draw the charts by hand, it is very tedious with Excel. What do you notice? Which presentation is preferable and why?

5.9 Create a pie chart for the variable education and interpret the result.

5.10 Create a bar chart for our variable motive, distinguish between service and industrial companies and interpret the result.

5.11 The second enterprise in our data set had a turnover of CHF 120,000 in 2007. In the first year it grew by 9%, in the second year by 15%, in the third year by 11%, in the fourth year by 9% and in the fifth year by -3%. Using this information, prepare a line graph showing the turnover development of the enterprise.

5.12 Search the Internet for the number of overnight stays for the Swiss holiday destinations Grison, Lucerne/Lake Lucerne, Berne upper country and Valais for the years 2005 to 2019 at the Swiss Federal Statistical Office. Plot the development of overnight stays using a line chart. We go into more detail, standardize the figures to the year 2008 (value 2008 equals 100) and redraw the line chart (this is already advanced). Describe briefly in a paragraph what is striking.

5.13 We are interested in the population density in South America and want to visualize it using a map chart. We go to the World Bank on the Internet and find the data for 2018. Then we prepare the data and map it using Excel.

Correlation: From Relationships

6

6.1 Correlation: The Joint Movement of Two Variables

Correlation, or the relationship between variables, is another central concept in statistics, along with averages, variation and frequencies. Here we are interested in whether variables are related in some way, for example the number of cheeseburgers people eat per week and the amount of weight they carry around their belly. There might be a correlation, let's think of Morgan Spurlock. In a self-experiment he ate only fast food for a while. Or let's think of the ex-mayor of New York Michael Bloomberg, who wanted to ban the sale of XXL soft drinks. In these cases, a correlation is assumed. The more cheeseburgers people eat, the larger are their bellies. But it could also be the other way around, the larger the bellies are, the more cheeseburgers they eat. The last statement gives an important piece of information. Correlation does not tell us the direction of the relationship or about causality. If we calculate a correlation, we don't know whether people who eat a lot of cheeseburgers weigh more or whether people who weigh more eat more cheeseburgers. All we know is that the variables move in the same direction.

Correlations can be positive or negative. This is not a judgment, it is only a statement about the direction of the correlation. Positive means that two variables move in the same direction. If the value of one variable becomes larger, so does the value of the other variable and vice versa. Negative means that two variables move in the opposite direction. If the values of one variable become larger, the values of the other variable become smaller and vice versa (Table 6.1).

Much of the knowledge we possess is based on correlation between variables. So, once again, it's worth looking at it. If we want to understand how correlations shape our knowledge, we need to understand correlation coefficients. Again, the correlation coefficients to be used depend on the scales. For metric data we use the correlation

Table 6.1 Correlation between two variables

Variable 1	Variable 2	Correlation	Example
Value of X increases	Value of Y increases	Positive correlation	Number of cheeseburgers per week becomes larger, amount of fat around the belly becomes larger
Value of X increases	Value of Y decreases	Negative correlation	Number of carrots per week becomes larger, amount of fat around the belly becomes smaller
Value of Y increases	Value of X increases	Positive correlation	Amount of fat around belly becomes larger, Number of cheeseburgers per week becomes larger
Value of Y increases	Value of X decreases	Negative correlation	Amount of fat around belly becomes larger, number of carrots per week becomes smaller

coefficient of Bravais–Pearson, for ordinal data the correlation coefficient of Spearman, and for nominal data the phi coefficient or the contingency coefficient.

6.2   The Correlation Coefficient of Bravais–Pearson for Metric Variables

The Bravais–Pearson correlation coefficient is used to determine the strength of a linear relationship between two metric variables. It is calculated as follows:

$$r = \frac{\sum (x_i - \overline{x})(y_i - \overline{y})}{\sqrt{\sum (x_i - \overline{x})^2 \sum (y_i - \overline{y})^2}}$$

We already know all the parts of the formula.

$\overline{x}$ and $\overline{y}$ are the arithmetic means and
x_i and y_i are the observed values,

i.e. we calculate the Bravais–Pearson correlation coefficient with the deviations of the observed values from their mean (concept of variation, compare Chap. 4).

Freak Knowledge

The upper part of the formula, the numerator, is the joint variance of the related x_i and y_i values from their respective mean, we also call it covariance σ_{xy}. If each of the deviations point in the same direction, are always positive or always negative, then the absolute value of the covariance becomes large and we have a correlation. If the deviations vary in an unsystematic way, the value becomes small and there is no

(continued)

correlation. The lower part of the formula, the denominator, is needed to standardize the Bravais–Pearson correlation coefficient to the range between -1 and 1.

How the correlation coefficient of Bravais–Pearson is calculated is shown with the help of the marketing and innovation variables of our data set. For example, we are interested in whether firms that spend a lot on marketing spend little on innovation and vice versa. Let's try this for the first six companies in the data set. The best way to start with is to create a table. The variable X should be our marketing variable, the variable Y the innovation variable. In fact which variable is X and which is Y does not matter.

Enterprise	x_i	y_i	$x_i - \bar{x}$	$(x_i - \bar{x})^2$	$y_i - \bar{y}$	$(y_i - \bar{y})^2$	$(x_i - \bar{x})(y_i - \bar{y})$
1	29	3	9	81	-3	9	-27
2	30	5	10	100	-1	1	-10
3	16	6	-4	16	0	0	0
4	22	5	2	4	-1	1	-2
5	9	9	-11	121	3	9	-33
6	14	8	-6	36	2	4	-12
$\bar{x}$	20	6					
$\sum$				358		24	-84

Inserting the three sums into our formula, we get the correlation coefficient of -0.91:

$$r = \frac{\sum (x_i - \bar{x})(y_i - \bar{y})}{\sqrt{\sum (x_i - \bar{x})^2 \sum (y_i - \bar{y})^2}} = \frac{-84}{\sqrt{358 \times 24}} = -0.91$$

How to interpret this value? First, we need to know that the correlation coefficient of Bravais–Pearson is defined in the range between -1 and 1. Values outside -1 and 1 are not possible. -1 means that we have a perfect negative correlation, 1 means a perfect positive correlation. To interpret values in between, the following rule of thumb is helpful (Table 6.2).

This means that we have a strong negative correlation between the two variables. Enterprises that spend a lot on marketing spend little on innovation or companies that spend a lot on innovation spend little on marketing.

Table 6.2 Rule of thumb to interpret the correlation coefficient

Value of r	Correlation between two variables
$r = 1$	Perfect positive correlation
$1 > r \geq 0.6$	Strong positive correlation
$0.6 > r \geq 0.3$	Weak positive correlation
$0.3 > r > -0.3$	No correlation
$-0.3 \geq r > -0.6$	Weak negative correlation
$-0.6 \geq r > -1$	Strong negative correlation
$r = -1$	Perfect negative correlation

6.3   The Scatterplot

We can graphically present the correlation in a scatterplot. A scatterplot is a dot diagram in which we plot our people or objects ordered by pairs of values. For our companies, we have the following pairs of values with respect to the variables marketing and innovation:

Enterprise	Marketing x_i	Innovation y_i	Pair of values $(x_i; y_i)$	Statement
1	29	3	29;3	Enterprise 1 spends 29% of sales on marketing and 3% on innovation
2	30	5	30;5	Enterprise 2 spends 30% and 5%
3	16	6	16;6	Enterprise 3 spends 16% and 6%
4	22	5	22;5	Enterprise 4 spends 22% and 5%
5	9	9	9;9	Enterprise 5 spends 9% and 9%
6	14	8	14;8	Enterprise 6 spends 14% and 8%

If we draw the enterprises based on the x_i and y_i values in an XY-diagram, we get our scatterplot (Fig. 6.1).

We see that enterprises that spend a lot on innovation spend little on marketing and vice versa. Furthermore, if we draw a straight line in the scatterplot, we get information about the strength of the linear relationship or how much the variables move together (Fig. 6.2).

If the points are close to the straight line, as in our case, there is a strong correlation. If the points vary widely around the straight line, as constructed in the following case, there is a weak correlation (Fig. 6.3).

In the scatterplot we can also see whether we are dealing with a positive or a negative correlation. If the straight line rises, there is a positive correlation, if the straight line falls, there is a negative correlation (Fig. 6.4).

In addition, we can see from the scatterplot whether there is any correlation at all between two variables. We have no correlation between two variables if the variables behave independently of each other. Graphically, this looks, for example, as in Fig. 6.5.

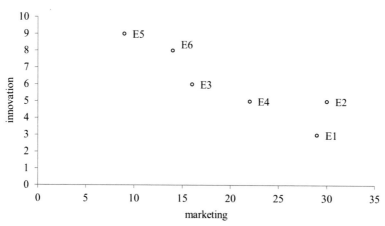

Fig. 6.1 Scatterplot for the variables marketing and innovation

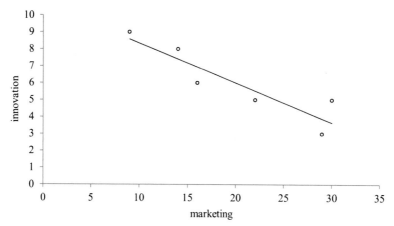

Fig. 6.2 Scatterplot for the variables marketing and innovation with strong relationship

In the left figure the x_i values increase, while the y_i values remain constant. In the middle figure the y_i values vary, while the x_i values do not change. In the right figure, both the x_i values and the y_i values vary. However, this variation is unsystematic, i.e. for large x_i values we have both small and large y_i values, for small x_i values we have both small and large y_i values.

The scatterplot is also important for another reason, namely when calculating the Bravais–Pearson correlation coefficient. It tells us whether there is a linear relationship between two variables. A linear relationship is the prerequisite for the Bravais–Pearson correlation coefficient to be calculated.

It should also be noted that the correlation coefficient of Bravais–Pearson can be strongly influenced by unusual observations or outliers. The scatterplot is here a helpful tool to detect them. Once we have detected such an observation in the scatterplot, we can

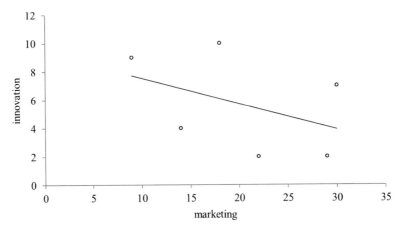

Fig. 6.3 Scatterplot for the variables marketing and innovation with weak relationship

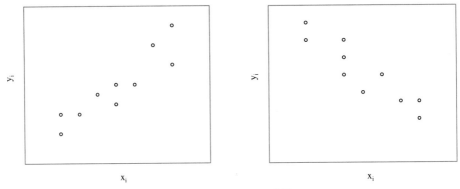

Fig. 6.4 Positive and negative correlation between two variables

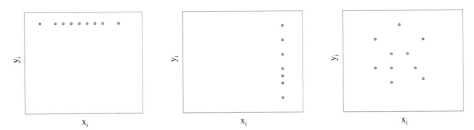

Fig. 6.5 No correlation between variables

treat it accordingly. We can calculate the correlation coefficient with and without outlier to determine the influence of the unusual observation. We can correct the unusual observation if it is a data entry error, remove it if it is irrelevant for our analysis, or keep the outlier if we consider it to be important.

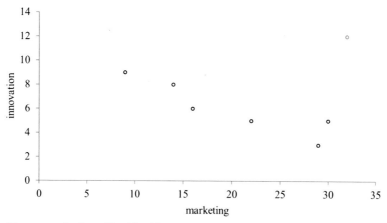

Fig. 6.6 The scatterplot from Fig. 6.2 with constructed outlier

As an example, we look again at the scatterplot from Fig. 6.2. However, we have added a constructed outlier (Fig. 6.6). If we now calculate the correlation coefficient without or with the outlier, we get completely different values.

For all the above reasons, we should always draw the scatterplot first before calculating the correlation coefficient of Bravais–Pearson. We see whether there is a linear relationship, we learn whether the relationship is positive or negative, we get an impression how strong the relationship is, and we can see if our relationship is affected by outliers.

> **Freak Knowledge**
> It happens that the relationship between the variables we are interested in is non-linear in nature, e.g. exponential. In such a case, we can often establish a linear relationship by mathematically transforming one of the two variables. Subsequently, the Bravais–Pearson correlation coefficient can be used again.

6.4   The Correlation Coefficient of Spearman for Ordinal Variables

If we have ordinal data instead of metric data, we use the rank correlation coefficient of Spearman. As with Bravais and Pearson, Spearman is the one who developed this coefficient. So there are real people and considerations behind our measures. It is calculated as follows:

$$r_{\mathrm{Sp}} = \frac{\sum \left(R_{x_i} - \overline{R}_x\right)\left(R_{y_i} - \overline{R}_y\right)}{\sqrt{\sum \left(R_{x_i} - \overline{R}_x\right)^2 \sum \left(R_{y_i} - \overline{R}_y\right)^2}}$$

where

R_{xi} and R_{yi} are the ranks of the observed x-values and y-values,
$\overline{R}_x$ and $\overline{R}_y$ are the mean values of the respective ranks.

We see, the rank correlation coefficient of Spearman is very similar to the correlation coefficient of Bravais–Pearson for metric data. The difference is that we have to give a rank to each observed value before we can calculate it. Typically, we give the largest value of a variable the rank of one and the smallest value the highest rank. If values occur more than once, we assign the middle rank for the values and then continue with the rank at which we would be without multiple occurrences. After we have assigned ranks for all values, we can calculate the required figures and insert them into the formula. The best way to explain this is to use the example of our first six companies. We are interested in whether there is a correlation between the two ordinal variables expectation X and self-assessment Y.

Enterprise	x_i	y_i	R_{x_i}	R_{y_i}	$R_{x_i} - \overline{R}_x$	$\left(R_{x_i} - \overline{R}_x\right)^2$	$R_{y_i} - \overline{R}_y$	$\left(R_{y_i} - \overline{R}_y\right)^2$	$\left(R_{x_i} - \overline{R}_x\right)\left(R_{y_i} - \overline{R}_y\right)$
1	2	5	3	1.5	−0.5	0.25	−2	4	1
2	1	4	5.5	3.5	2	4	0	0	0
3	3	5	1	1.5	−2.5	6.25	−2	4	5
4	2	3	3	5.5	−0.5	0.25	2	4	−1
5	1	3	5.5	5.5	2	4	2	4	4
6	2	4	3	3.5	−0.5	0.25	0	0	0
$\overline{R}$			3.5	3.5					
$\sum$						15		16	9

For the variable X, we look for the largest value and give it a rank of 1. The largest value is 3, which we give the rank of 1. We then look for the second largest value. This is the number 2, it occurs three times, i.e. for enterprise 1, 4 and 6. The enterprises would therefore receive the rank 2 if we were to assign it more often. However, we always assign a rank only once. Since the enterprises 1, 4 and 6 have equal values, we give them the middle rank. It is calculated by dividing the sum of the ranks by the number of ranks to give. We have three ranks to assign, rank 2, 3 and 4 for the three enterprises, the sum is hence 9, we divide by 3 ranks, i.e. the rank to assign is 3. Now we have assigned the first four ranks, we continue with the remaining companies 2 and 5 and give them the middle rank from ranks 5 and 6, i.e. rank 5.5. We proceed in the same way with the variable Y. Then we calculate the values needed for the formula and put them into the formula. Now we can calculate the rank correlation coefficient o Spearmen, which is:

$$r_{Sp} = \frac{\sum (R_{x_i} - \overline{R}_x)(R_{y_i} - \overline{R}_y)}{\sqrt{\sum (R_{x_i} - \overline{R}_x)^2 \sum (R_{y_i} - \overline{R}_y)^2}} = \frac{9}{\sqrt{15 \times 16}} = 0.58$$

We interpret the rank correlation coefficient in the same way as the Bravais–Pearson correlation coefficient. Thus, we have found a positive correlation between the two variables. If the value for the variable X becomes larger, the value of the variable Y becomes larger and vice versa. In our case, this means that less experienced people tend to have a slightly worse assessment of future developments and vice versa. The interpretation is a bit tricky. To do the interpretation, you have to look very carefully in the legend to see how the data is defined.

Freak Knowledge

If no values occur twice within our respective variables, i.e. in variable X as well as variable Y, we can simplify the formula used above. The formula for Spearman's rank correlation coefficient is then

$$r_{sp} = 1 - \frac{6 \times \sum d_i^2}{n^3 - n}$$

with n for the number of observations, and d_i for the distances between the ranks.

6.5   The Phi Coefficient for Nominal Variables with Two Characteristics

If we have nominal variables with only two characteristics, e.g. the variable gender with the values man and woman or 0 and 1, the correlation can be calculated using the phi coefficient:

$$r_\phi = \frac{a \times d - b \times c}{\sqrt{S_1 \times S_2 \times S_3 \times S_4}}$$

where

r_ϕ	is the phi coefficient,
a, b, c, d	are the frequencies of the fields of a 2×2 matrix, and
S_1, S_2, S_3, S_4	are the sums of the rows and columns.

		Variable Y		
Variable X		0	1	
	0	a	b	S_1
	1	c	d	S_2
		S_3	S_4	

Let's look at this using the first ten enterprises of our data set data_growth.xlsx as an example. Nominal variables with only two characteristics are sex and sector. Sex distinguishes between female and male founders, and sector between industry and service enterprises. We want to know whether there is a relationship between the sex of the founders and the sector in which the enterprise was founded. The data show that the first enterprise is an industry firm and was founded by a man, the second firm is a service firm founded by a woman, the third is again an industry firm, founded by a man, and so on. (compare our data set for the first ten enterprises). To summarize, we can present the information in the following 2×2 *matrix*:

		Sex		
		0 = Man	1 = Woman	
Sector	0 = Industry	2	0	2
	1 = Service	3	5	8
		5	5	

If we put the numbers into our formula accordingly, we get the phi coefficient:

$$r_\phi = \frac{a \times d - b \times c}{\sqrt{S_1 \times S_2 \times S_3 \times S_4}} = \frac{2 \times 5 - 0 \times 3}{\sqrt{2 \times 8 \times 5 \times 5}} = 0.50$$

We can interpret the phi coefficient almost like the correlation coefficients of Bravais–Pearson and Spearman. The phi coefficient is also defined between -1 and 1. The only difference is that the sign is meaningless. We could easily check this, when we, for example, swap the male and female columns, the sign becomes then negative. Hence, we only interpret the absolute value and find that we have a medium relationship between the two variables. Now, if we want to know how the relationship is, we need to look again at our 2×2 *matrix*. We see that relatively many service firms were founded by women and that women founded relatively few industrial firms. This also allows us to determine the tendency. Women tend to be more likely to start service enterprises.

6.6   The Contingency Coefficient for Nominal Variables

If we are faced with nominal data with more than two characteristics, it is enough if one of the nominal data has more than two, then the correlation can be calculated using the contingency coefficient as follows:

$$C = \sqrt{\frac{U}{U + n}}$$

with

C is the contingency coefficient,
n is the number of observations, and
U is the sum of the deviations between the observed and theoretically expected values:

$$U = \sum \sum \frac{\left(f_{jk} - e_{jk}\right)^2}{e_{jk}}$$

where

f_{jk} are the observed frequencies in the cells,
e_{jk} are the theoretically expected frequencies in the cells,
j are the rows and
k are the columns.

Defined is the contingency coefficient not between -1 and 1, but in the range from 0 to a maximum: $0 \le C \le C_{Maximum}$. $C_{Maximum}$ is calculated from the number of columns and the number of rows, it is always less than 1:

$$C_{maximum} = \frac{1}{2} \left(\sqrt{\frac{r - 1}{r}} + \sqrt{\frac{c - 1}{c}} \right)$$

with

r is equal to the number of rows and
c is equal to the number of columns.

At this point, it is useful to start directly with an example to clarify the issues said. Let's take our data set data_growth.xlsx again and analyze the relationship between the variables motive and sector. This time we use all of the 100 enterprises observed, since none of the cells should be occupied with less than five frequencies. According to the phi coefficient, we can construct a matrix, but in this case with two rows and three columns, since the variable sector has two characteristics and the variable motive has three. Theoretically, the matrix has the following structure with two rows and three columns:

Rows	Columns		
	1	2	3
1	f_{11}	f_{12}	f_{13}
2	f_{21}	f_{22}	f_{23}

In our example, it looks like this, with the counted observed frequencies.

Sector	Motive		
	Unemployment	Implement idea	Higher income
Industry	8	15	11
Service	9	39	18

The table shows that eight industry foundations from male founders are founded out of unemployment, 15 male founders founded the industry enterprise because of an idea, and so on. We have two rows and three columns, so we can furthermore calculate the in maximum achievable contingency coefficient:

$$C_{\text{maximum}} = \frac{1}{2}\left(\sqrt{\frac{r-1}{r}} + \sqrt{\frac{c-1}{c}}\right) = \frac{1}{2}\left(\sqrt{\frac{2-1}{2}} + \sqrt{\frac{3-1}{3}}\right) = 0.76$$

At most, we can achieve a contingency coefficient of 0.76. If we discover such a value, then we have a perfect correlation. Minimally we can achieve 0, we would have no correlation between sector and founding motive.

The next step is to calculate the theoretically expected values. To accomplish this, we first need to calculate the row and the column sums.

Sector	Motive			Total
	Unemployment	Implement idea	Higher income	
Industry	8	15	11	34
Service	9	39	18	66
Total	17	54	29	

To get the theoretically expected values, we use the following formula:

$$e_{jk} = \frac{f_j \times f_k}{n}$$

with

f_j are the row sums and
f_k are the column sums.

This leads to the following matrix:

Sector	Motive			Total
	Unemployment	Implement idea	Higher income	
Industry	$e_{11} = \frac{34 \times 17}{100} =$ 5.78	$e_{12} = \frac{34 \times 54}{100} =$ 18.36	$e_{13} = \frac{34 \times 29}{100} =$ 9.86	34
Service	$e_{21} = \frac{66 \times 17}{100} =$ 11.22	$e_{22} = \frac{66 \times 54}{100} =$ 35.64	$e_{23} = \frac{66 \times 29}{100} =$ 19.14	66
Total	17	54	29	

We notice that the row and column totals have remained the same. However, the cells now contain the values that we would theoretically expect if the respective row adjusted to the ratio of the column totals.

This allows us to calculate the deviation sum U by subtracting the observed values from the theoretically expected values:

$$U = \sum \frac{\left(f_{jk} - e_{jk}\right)^2}{e_{jk}}$$

$$= \frac{(f_{11} - e_{11})^2}{e_{11}} + \frac{(f_{12} - e_{12})^2}{e_{12}} + \frac{(f_{13} - e_{13})^2}{e_{13}} + \frac{(f_{21} - e_{21})^2}{e_{21}} + \frac{(f_{22} - e_{22})^2}{e_{22}} + \frac{(f_{23} - e_{23})^2}{e_{23}}$$

$$= \frac{(8 - 5.78)^2}{5.78} + \frac{(15 - 18.36)^2}{18.36} + \frac{(11 - 9.86)^2}{9.86} + \frac{(9 - 11.22)^2}{11.22} + \frac{(39 - 35.64)^2}{35.64} + \frac{(18 - 19.14)^2}{19.14}$$

$$= 2.42$$

If we put the deviation sum into the formula for the contingency coefficient, we get the coefficient:

$$C = \sqrt{\frac{U}{U + n}} = \sqrt{\frac{2.42}{2.42 + 100}} = 0.15$$

That is, we are closer to zero than to the maximum possible contingency coefficient, so we have no correlation between the variables motive and sector.

6.7   Correlation, Spurious Correlation, Causality, and More Correlation Coefficients

Be careful, when we calculate a correlation, we know nothing about causality, i.e. we don't know whether

X $\longrightarrow$ Y   variable X influences variable Y,

Y $\longrightarrow$ X   variable Y influences variable X,

X $\longleftrightarrow$ Y   variable X and variable Y mutually influence each other.

Let's look at this again with an example. Between the variables marketing and innovation we calculated a very high negative correlation of -0.91. It follows that enterprises that spend a lot of their turnover on marketing spend little of their turnover on innovation. However, we do not know which way the relationship runs, for example, whether enterprises that spend a lot on marketing have little money left over for innovation, or the other way around, enterprises that spend a lot on innovation have little money left over for marketing. Perhaps there is a reciprocal relationship. If a little more money is spent on marketing, less is left over for innovation. If less money is then spent on innovation, more is in turn spent on marketing, and so on.

Causality is a matter of theory or theoretical reasoning. We have knowledge about causality when we can clearly detect from theory that a variable X influences a variable Y. Then and only then do we know something about causality. An example of this is a stone falling into water. We hold a stone, release it, it falls into the water, this causes waves on the water. The waves on the surface of the water are a direct result of the stone. Once again, correlation says nothing about causality.

One more thing is important. Many correlations are caused by a third variable, a variable that is behind the correlation between variables X and Y.

In Fig. 6.7, such a situation is illustrated. There is a correlation between the variables X and Y, but it is not based on a correlation between X and Y, but variable X is correlated with variable Z and variable Y as well. The correlations between X with Z and Y with Z are behind the correlation between X and Y. We speak of a spurious correlation. A classic example of a spurious correlation is the positive correlation between the number of storks and the birth rate observed during the period of industrialization. The number of storks declined and so did the birth rate. This does not mean that there is a correlation between storks and births. The variable behind this is industrialization. Industrialization caused individuals to become richer and having fewer children. At the same time, the accompanying increase in pollution has reduced the number of storks. Another example

Fig. 6.7 Spurious correlation

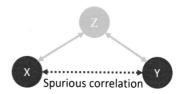

Spurious correlation

Table 6.3 Correlation coefficients and scales of variables

Variable X	Variable Y	Correlation coefficient	Example from the data set data_growth.xlsx
Metric	Metric	Bravais–Pearson Correlation Coefficient	Correlation between growth rate and marketing
Metric	Ordinal	Spearman Rank Correlation Coefficient	Correlation between growth rate and self-assessment
Metric	Nominal	Point biserial correlation	Correlation between growth rate and sex
Ordinal	Ordinal	Spearman rank correlation coefficient	Correlation between self-assessment and education
Ordinal	Nominal	Biserial rank correlation	Correlation between self-assessment and sex
Nominal	Nominal	Contingency Coefficient	Correlation between motive and sex
Nominal (0/1)	Nominal (0/1)	Phi coefficient	Correlation between sector and sex

is a positive correlation between the amount of ice cream consumed each summer and the number of people who drown in a summer. The more ice consumed, the more people drown and vice versa. This does not mean that eating ice cream increases the risk of drowning or "that drowning victims consume more ice cream." What's behind this is the quality of the summer. If it's hotter, people eat more ice cream but also swim more and accordingly there are more deaths through drowning. So when we discover a correlation between two variables, we have to check very carefully to make sure we haven't found a spurious correlation. This brings us back to theory. Only theory tells if it is a true correlation.

If we calculate a correlation, we use Bravais–Pearson's correlation coefficient for metric data, Spearman's rank correlation coefficient for ordinal data, and either the phi coefficient or the contingency coefficient for nominal data. But what do we do when we have mixed data, for example, when one variable is metric and the other is ordinal? If one variable is metric and the other is ordinal, then we can use Spearman's rank correlation coefficient. If one variable is metric and the other is nominal, then we need to calculate the point biserial correlation. The only remaining situation is when one variable is ordinal and the other is nominal. In such a case, we use the biserial rank correlation (Table 6.3).

6.8   Calculating Correlation Coefficients with Excel

6.8.1   Calculating the Correlation Coefficient Bravais–Pearson with Excel Using the Command Insert Function

To calculate the Bravais–Pearson correlation coefficient, we can use the function CORREL with the Formula tab. We select the command CORREL under Insert Function and enter the range of values for the first variable at Array1 and the range of values for the second variable at Array2.

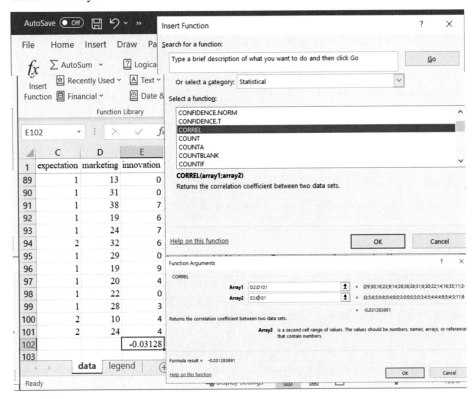

6.8.2   Calculating the Correlation Coefficient Bravais–Pearson with Excel Using the Data Analysis Command

The Bravais–Pearson correlation coefficient can also be calculated under the Data tab with the Data Analysis command. Within the Data Analysis window, we select the command

Correlation and enter the range of values (the variables should be in adjacent columns). We have further to define the output range to obtain the correlation coefficients.

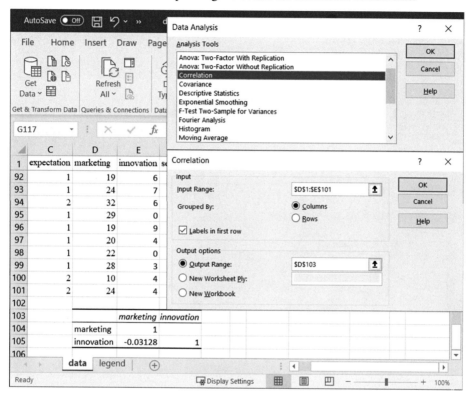

6.8.3   Determine the Ranks for an Ordinal Variable with Excel

There is no function in Excel to calculate Spearman's rank correlation coefficient. However, Excel can be used to determine the ranks. To do this, we select the function RANK.AVG under the Formulas tab. Using this function, we first determine the ranks, then calculate the required values and insert them into our formula for Spearman's rank correlation coefficient. In the window, we see the ranks for the variable expectation.

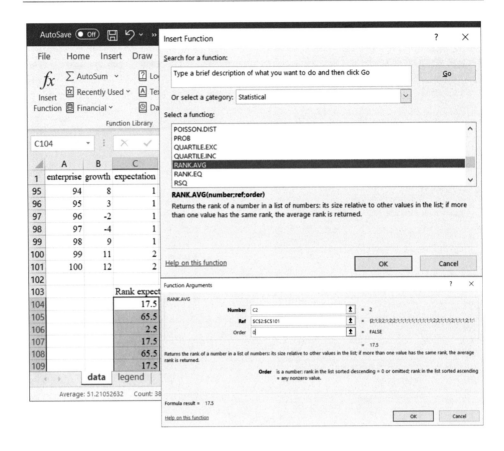

6.8.4 Creating a Pivot Table with Excel

Also missing is a function to calculate the phi coefficient or the contingency coefficient. To make our job easier, we can create a table of observed frequencies using the PivotTable button under the Insert tab. Once we have done this, proceed as described earlier in the respective sections.

To create a pivot table with Excel, it is a good idea first to select the entire range of values.

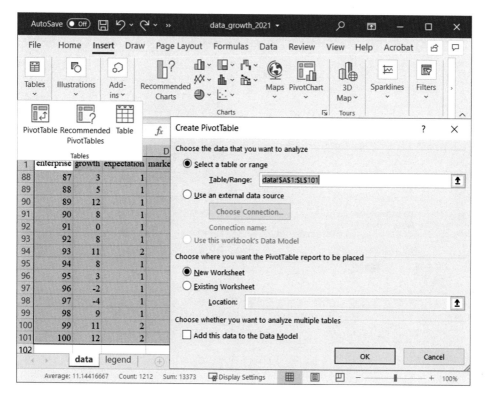

Then we click on PivotTable under the Insert tab and in the Create PivotTable window we define best that the pivot table is created in a new worksheet. We click OK and get the following window:

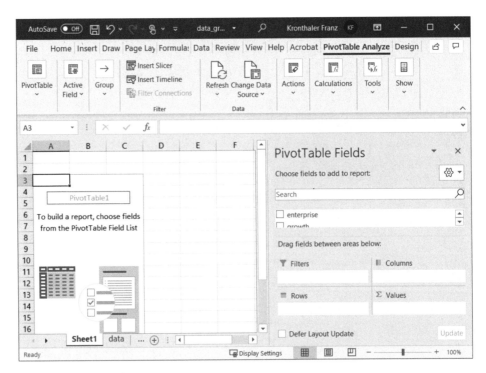

In this window, we click on the variables for which we want to create the crosstab. In our case, these are the variables sector and motive (we used this example above with the contingency coefficient). Then we shape the table by dragging and clicking until we get a crosstab (at this point I refer to the help function of Excel).

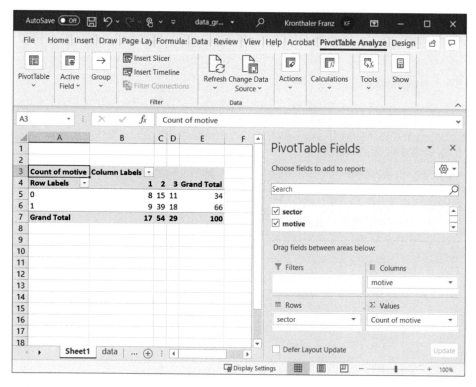

At the end we need to calculate the contingency coefficient using the obtained observed values.

6.9   Checkpoints

- *Correlations provide information about whether there is a relationship between two variables, they do not say anything about causality.*
- *It is often the case that correlations are due to the influence of third variables. Correlations must always be theoretically justified.*
- *The Bravais–Pearson correlation coefficient is suitable for metric variables, when a linear relationship exists.*
- *Before calculating the Bravais–Pearson correlation coefficient, the scatterplot should be drawn.*
- *The correlation coefficient of Bravais–Pearson is sensitive to unusual observations or outliers.*
- *The Spearman correlation coefficient is used to calculate the correlation between two ordinal variables.*
- *If a correlation between two nominal variables is analyzed the phi coefficient or the contingency coefficient should be used.*

- *Other specialized correlation coefficients are available for calculating correlations between variables with different scales.*

6.10 Applications

6.1 Which correlation coefficient should be used for which scales?

6.2 Why are theoretical considerations necessary before calculating a correlation between variables?

6.3 Why does a correlation coefficient say nothing about causality?

6.4 Consider the following three scatterplots, which were discussed by Anscombe nearly fifty years ago in the year 1973. In which of the cases can the Bravais–Pearson correlation coefficient be calculated without problems? Why?

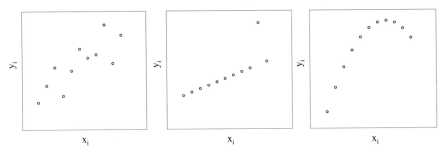

6.5 Draw and calculate by hand the scatterplot and the correlation coefficient for the variable growth and marketing for the first eight companies of our data set data_growth.xlsx and interpret the result.

6.6 Use Excel to draw the scatterplot and to calculate the correlation coefficient of Bravais–Pearson for the variables growth and marketing, growth and innovation, growth and experience, and growth and age and interpret the result. The data set is data_growth.xlsx

6.7 Calculate by hand the rank correlation coefficient of Spearman for the variables growth rate and self-assessment and interpret the result, use the first ten companies in the data set data_growth.xlsx.

6.8 Use Excel to calculate the rank correlation coefficient of Spearman for the variables growth and self-assessment and interpret the result. The data set is data_growth.xlsx.

6.9 Calculate the phi coefficient for the variables sex and sector and interpret the result, data set data_growth.xlsx.

6.10 For our data set data_growth.xlsx, calculate the contingency coefficient for the variables sex and motive and interpret the result.

6.11 We are interested in the relationship between CO_2 emissions per capita and the gross domestic product per capita in the European Union. In the database of Eurostat (https://ec.europa.eu/eurostat/de/home) we find the following information:
Calculate and interpret the correlation coefficient.

Country	GDP per capita 2017 [in 1000 €/per capita]	CO_2 per capita 2017 [in t/per capita]
Belgium	35.3	9.0
Bulgaria	6.3	6.8
Czech Republic	17.2	10.2
Denmark	47.4	6.6
Germany	35.4	10.0
Estonia	14.4	14.3
Ireland	54.2	8.7
...		

6.12 We are interested in whether there is a correlation between three stocks. We once again go to the Internet, visit a financial platform (this time perhaps https://www.boerse-online.de/) and retrieve historical values for VW, Daimler and SAP. We get the following values:

Date	VW	Daimler	SAP
01.04.2020	99.27	25.74	97.40
31.03.2020	105.48	27.14	101.48
30.03.2020	105.50	27.21	101.70
27.03.2020	106.50	27.32	100.78
26.03.2020	112.90	29.21	103.88
25.03.2020	115.50	30.25	100.00

Calculate the possible correlations between the stocks and interpret the results. Why might we are interested?

Ratios and Indices: The Opportunity to Generate New Knowledge from Old Ones

7

So far, we have analyzed variables that are simply available. But what if we do not have the information yet, but it is hidden in our data. We can then generate the information from our existing variables by relating or weighting them. Let's look at the profit of an enterprise. It can make a difference whether we measure the profit in absolute terms or in relative terms. Let's assume that we have a small enterprise with ten employees and an enterprise with 100 employees. Both enterprises should make a profit of CHF 500'000 per year. In absolute terms, the profit is the same. But in relative terms, per employee, it is CHF 50'000 per employee for the smaller company and only CHF 5'000 per employee for the larger company. We have formed a ratio that allows us to compare the enterprises:

$$\text{Profit per capita} = \frac{\text{Profit in CHF}}{\text{Employee}}$$

Ratios are quotients of two numbers. They can be used to compare persons and other objects in terms of content, space or time. In the following we present different ratio numbers and discuss their application. We will not refer to our data set, but show different situations where ratio numbers are used. Furthermore, we will discuss index numbers, which we use when price or quantity developments of several products are to be analyzed simultaneously.

© Springer-Verlag GmbH Germany, part of Springer Nature 2023
F. Kronthaler, *Statistics Applied With Excel*,
https://doi.org/10.1007/978-3-662-64319-8_7

Table 7.1 Gross domestic product per capita of selected countries—comparison using a reference number (World Bank 2020, data downloaded from https://data.worldbank.org/ on 10.04.2020)

Country	GDP (in $, 2018)	Population (2018)	GDP per capita (in $, 2018)
China	13'608'151'864'638	1'392'730'000	9'771
Germany	3'947'620'162'503	82'927'922	47'603
South Africa	368'288'939'768	57'779'622	6'374
Switzerland	705'140'354'166	8'516'543	82'797
USA	20'544'343'456'937	327'167'434	62'795

7.1 Different Ratio Numbers

First we can form ratios by dividing two different parameters, we speak sometimes of forming a reference number:

$$\text{Reference number} = \frac{\text{Parameter A}}{\text{Parameter B}}$$

If it makes sense, we multiply the number by 100 to get a percentage value.

Probably one of the most familiar reference number is the gross domestic product per capita. It is formed by dividing the gross domestic product of a country by the population of that country. This makes it possible to compare the average wealth of the population in different countries. Table 7.1 displays the gross domestic product, the population and the gross domestic product per capita of selected countries.

In absolute terms, the USA appears to be the richest nation in the table, followed by China, Germany, Switzerland and South Africa. Relative to the population, the picture is different, with the Swiss population the richest, followed with a large difference by the USA, Germany, China and South Africa. Depending on whether we operate with absolute or relative numbers, a different picture shows up.

Other examples of reference number are

- the car density, number of registered vehicles to the population,
- the birth rate, the number of live-births to the population,
- the physician density, number of physicians to the population or to the area of a country,
- the return on equity as profit divided by the equity,
- the productivity, as the labor output to the hours worked, or...
- beer consumption per capita.

There are many ways to form a reference numbers, and it depends on the question being asked, which two numbers are used. For example, if we want to make a statement about the physician density of a country, we have to consider whether it makes sense to calculate

Table 7.2 Export ratio of selected countries – comparison using a quota (World Bank 2020, data downloaded from https://data.worldbank.org/ on 09.04.2020)

Country	Exports (in $, 2018)	GDP (in $, 2018)	Exports/GDP (in %, 2018)
China	2'655'609'104'875	13'608'151'864'638	19.5
Germany	1'871'807'732'127	3'947'620'162'503	47.4
South Africa	110'144'477'441	368'288'939'768	29.9
Switzerland	466'305'408'689	705'140'354'166	66.1
USA	2'510'250'000'000	20'544'343'456'937	12.2

the number of physicians per area or per population. Depending on which approach we choose, we will probably come to different conclusions.

A second way to form a ratio number is to divide a share of a parameter by its total parameter, sometimes this is called a quota:

$$\text{Quota} = \frac{\text{Share parameter}}{\text{Total parameter}}$$

Usually we multiply the number by 100 and we get the percentage share of the part being used to its total.

In Table 7.2, we see the export ratio of selected countries as an example.

The export ratio is the ratio of the value of goods exported to the gross domestic product. The figure illustrates which country exports more of the goods produced in the country. We see that the export ratio is highest in Switzerland, followed by Germany. The export ratio of the United States is the smallest. One of the reasons for this is that the U.S. has a large domestic market and therefore companies can sell a relatively large number of the goods produced at home. Thus, they are less dependent on export markets.

Other examples of quotas are

- the odds having a boy, number of newborn boys to the total number of newborns,
- the export ratio of a firm, export sales to total sales,
- the national savings rate, national saving to gross domestic product,
- the equity ratio of a company, equity to its capital,
- the marketing ratio, marketing expenditure to total expenditure.

Again, we can see that there are many ways to form quotas and that there are other ratios depending on the question.

The last ratio number we would like to discuss is the dynamic index. We use this to look at and compare developments over time:

$$\text{Dynamic index}_t = \frac{\text{Value of the parameter at time point t}}{\text{Value of the parameter at the starting point}}$$

Table 7.3 Population growth of selected countries—comparison using the dynamic index figure (World Bank 2020, data downloaded from https://data.worldbank.org/ on 09.04.2020)

Year	China		Germany		Switzerland	
	Population	I_{2005}	Population	I_{2005}	Population	I_{2005}
2005	1'303'720'000	100.0	82'469'422.0	100.0	7'437'115	100.0
2006	1'311'020'000	100.6	82'376'451.0	99.9	7'483'934	100.6
2007	1'317'885'000	101.1	82'266'372.0	99.8	7'551'117	101.5
2008	1'324'655'000	101.6	82'110'097.0	99.6	7'647'675	102.8
2009	1'331'260'000	102.1	81'902'307.0	99.3	7'743'831	104.1
2010	1'337'705'000	102.6	81'776'930.0	99.2	7'824'909	105.2
2011	1'344'130'000	103.1	80'274'983.0	97.3	7'912'398	106.4
2012	1'350'695'000	103.6	80'425'823.0	97.5	7'996'861	107.5
2013	1'357'380'000	104.1	80'645'605.0	97.8	8'089'346	108.8
2014	1'364'270'000	104.6	80'982'500.0	98.2	8'188'649	110.1
2015	1'371'220'000	105.2	81'686'611.0	99.1	8'282'396	111.4
2016	1'378'665'000	105.7	82'348'669.0	99.9	8'373'338	112.6
2017	1'386'395'000	106.3	82'657'002.0	100.2	8'451'840	113.6
2018	1'392'730'000	106.8	82'927'922.0	100.6	8'516'543	114.5

We also usually multiply this number by 100 to get the change to the basis we have chosen, as a percentage.

In Table 7.3 we see the population development of China, Germany and Switzerland measured in absolute terms and using the dynamic index at the basis of the year 2005.

Using the dynamic index, it is easy to compare the population trends of the three countries. We see that since 2005, the population in China has increased by 6.8% compared to the base year, the population in Germany has remained almost the same with a growth of 0.6%, and the population in Switzerland has increased by 14.5%. Switzerland thus recorded the highest population growth in relative terms during this period. In the same way, other time series can be compared using the dynamic index, e.g. the economic growth of countries, the growth of various companies, the development of shares and much more.

7.2   The Price and Quantity Index of Laspeyres and Paasche

If we want to analyze the development of one quantity, the dynamic index is a good thing. But let's imagine that we want to look at the development of several variables, e.g. a basket of goods or several shares, as e.g. it is done with a share index. The dynamic index is then no longer suitable. We need a number that represents the development of several related variables over time. Such numbers are available to us with the index of Laspeyres and Paasche.

Let's briefly consider this using the development of the cost of living as an example. To calculate the development of the cost of living, we could take the prices of the various goods, calculate the average prices for several points in time, e.g. years, and then use them to show the price development. However, this does not consider the importance of the individual goods. Let's think of our own basket of goods, which we buy weekly, monthly or annually. Some goods, such as bread, we buy almost daily, while other goods, such as a television or a car is less often budget relevant. Therefore, to determine the importance of the price trend of goods to our budget, we need to weight the goods according to their importance. It makes a difference whether petrol becomes 10% more expensive or a television.

This has been noticed before and a man called Laspeyres developed a price index in the 19th century. He took into account the importance of goods in the price trend. This index is still used today, as already indicated, for example, in the calculation of the development of consumer prices or in the calculation of share indices.

The price index according to Laspeyres is calculated with the following formula:

$$P_{Lt} = \frac{\sum p_{ti} \times q_{0i}}{\sum p_{0i} \times q_{0i}} \times 100$$

where

P_{Lt} denotes the Laspeyres price index at time t,
p_{ti} is the price of good i at time t,
p_{0i} is the price of good i at the base time 0,
q_{0i} is the quantity of good i at base time 0.

We demonstrate the calculation with a small specific example. Suppose a cafe owner wants to adjust the prices of the breakfast menu after five years. To make the adjustment, the owner looks at the quantities purchased and the prices today and five years ago. For simplicity, let's assume that breakfast consists only of coffee, bread, cheese, and ham. The numbers might look like this:

Goods	Price 2014	Quantity 2014	Price 2019	Quantity 2019
Coffee	5 CHF/kg	150 kg	7 CHF/kg	150 kg
Bread	2 CHF/piece	5'000 pieces	3 CHF/piece	6'000 pieces
Cheese	15 CHF/kg	500 kg	17 CHF/kg	750 kg
Ham	25 CHF/kg	250 kg	35 CHF/kg	150 kg

We insert the figures into the Laspeyres price index formula to obtain the price trend since 2014.

Compared to the base year 2014, prices have increased by 35.9 % until 2019 (135.9–100). So we have to pay 35.9% more for the goods, which means maybe we should adjust our breakfast prices upwards accordingly.

$$P_{L2019} = \frac{\sum p_{2019i} \times q_{2014i}}{\sum p_{2014i} \times q_{2014i}} \times 100 =$$

$$= \frac{7\frac{CHF}{kg} \times 150kg + 3\frac{CHF}{St\ddot{u}ck} \times 5000St\ddot{u}ck + 17\frac{CHF}{kg} \times 500kg + 35\frac{CHF}{kg} \times 250kg}{5\frac{CHF}{kg} \times 150kg + 2\frac{CHF}{St\ddot{u}ck} \times 5000St\ddot{u}ck + 15\frac{CHF}{kg} \times 500kg + 25\frac{CHF}{kg} \times 250kg} \times 100 =$$

$$= 135.9$$

For this, first, it doesn't matter in what units the goods are measured, the units are canceled out of the formula. Second, notice that we used the base year quantities in the numerator and the denominator. The quantities of the current year are not included. It follows that the Laspeyres price index shows how much more we would have to pay today for the basket of goods from the base year. This works well if the basket of goods has not changed much, but if the basket of goods is different today, then the Laspeyres price index is inaccurate.

Mr. Paasche has also noticed this. He suggested not calculating the price development with quantities from the base year, but with the quantities of the actual year t. The formula adjusts accordingly. The formula is hence

$$P_{Pt} = \frac{\sum p_{ti} \times q_{ti}}{\sum p_{0i} \times q_{ti}} \times 100$$

where

P_{Pt} denotes the Paasche price index at time t,

p_{ti} is the price of good i at time t,

p_{0i} is the price of good i at the base time 0,

q_{ti} is the quantity of good i at time t.

If we now insert the values accordingly, we get a different value for the price trend.

According to the Paasche price index, the basket of goods purchased today costs 33.5% (133.5–100) more than at the base time. The advantage of the Paasche price index is that

$$P_{P2019} = \frac{\sum p_{2019i} \times q_{2019i}}{\sum p_{2014i} \times q_{2019i}} \times 100 =$$

$$= \frac{7\frac{CHF}{kg} \times 150kg + 3\frac{CHF}{St\ddot{u}ck} \times 6000St\ddot{u}ck + 17\frac{CHF}{kg} \times 750kg + 35\frac{CHF}{kg} \times 150kg}{5\frac{CHF}{kg} \times 150kg + 2\frac{CHF}{St\ddot{u}ck} \times 6000St\ddot{u}ck + 15\frac{CHF}{kg} \times 750kg + 25\frac{CHF}{kg} \times 150kg} \times 100 =$$

$$= 133.5$$

it takes into account the situation of today. The disadvantage is that the values cannot be compared over time because the quantities are adjusted at each time.

The advantage of the Laspeyres price index is hence that it is comparable over time. We observe the price development of a basket of goods, with quantities remaining the same. The disadvantage is that the weighting scheme (the quantities) slowly become obsolete. The Laspeyres price index leads to results that become more irrelevant the more the weighting scheme is outdated. The advantage of the Paasche price index, on the other hand, is that it always uses the current weighting scheme. However, it is not comparable over time because the weighting scheme is always changing.

If comparability over time is important, then the Laspeyres price index should be used. From time to time, however, the weighting scheme should be adjusted, especially if it has changed significantly.

Just like the price development, the quantity development can be calculated with the Laspeyres index and the Paasche index. The only difference is that the importance of the goods is expressed by their prices, i.e. the weighting scheme is the price. Otherwise, the formulas are constructed identically except for the difference that now the quantities and not the prices change.

The formula for the Laspeyres quantity index is:

$$Q_{Lt} = \frac{\sum q_{ti} \times p_{0i}}{\sum q_{0i} \times p_{0i}} \times 100$$

The formula for Paasche's quantity index is:

$$Q_{Pt} = \frac{\sum q_{ti} \times p_{ti}}{\sum q_{0i} \times p_{ti}} \times 100$$

where

Q_{Lt} is the Laspeyres quantity index at time t, and
Q_{Pt} is the Paasche quantity index at time t.

The procedure of calculation and interpretation is identical and we omit it here. In general, however, it can be said that quantity indices are used when the quantitative development is of interest, e.g. at the company level, the development of the quantities of all products produced or at the macroeconomic level, the development of the production of industrial companies.

7.3   Checkpoints

- *Ratios relate two parameters to each other. We can use them to generate new knowledge from existing data.*
- *Reference numbers relate different quantities to each other.*
- *Quotas relate sub-quantities to an overall quantity.*
- *We can use a dynamic index to analyze the development of a variable over time.*
- Price and quantity indices by Laspeyres and Paasche are used when analyzing the development of several variables over time.

7.4   Applications

7.1 Find three facts that can be analyzed with the help of a reference number.

7.2 Find three facts that can be analysed using a quota.

7.3 Find three facts that can be analyzed using a dynamic index figure.

7.4 Analyze the development of the gross domestic product per capita for the following countries. Use 2007 as the base year.

GDP per capita (in $, current prices)

Year	China	Germany	Switzerland
2007	2'694	41'587	63'555
2008	3'468	45'427	72'488
2009	3'832	41'486	69'927
2010	4'550	41'532	74'606
2011	5'618	46'645	88'416
2012	6'317	43'858	83'538
2013	7'051	46'286	85'112
2014	7'651	47'960	86'606
2015	8'033	41'140	82'082
2016	8'079	42'099	80'172
2017	8'759	44'240	80'450
2018	9'771	47'603	82'797

Source: World Bank 2020, data downloaded from https://data.worldbank.org/ 11.04.2020.

7.5 A company needs energy from oil, gas, and electricity to produce its goods. The company wants to analyze the price and quantity trends of the energy needed for production. In particular, it wants to know the effect of efforts to produce in a more energy-efficient way. Calculate and analyze the Laspeyres and Paasche price and quantity indices.

Energy source	Price 2015	Quantity 2015	Price 2019	Quantity 2019
Oil	0.12 [CHF/liter]	70,000 [liters]	0.30 [CHF/liter]	64,000 [liters]
Gas	0.28 [CHF/m3]	10,000 [m3]	0.42 [CHF/m3]	14,000 [m3]
Electricity	0.08 [CHF/kWh]	280,000 [kWh]	0.06 [CHF/kWh]	220,000 [kWh]

7.6 As a student, you fortunately have a small fortune and have invested it in shares. Of course, you are interested in the performance of your portfolio and you would like to analyze the development, similar to the Swiss Market Index SMI or the German stock index DAX, with the help of an index value. With the help of the Laspeyres price index, develop an index for the stock portfolio that represents the price development.

	Share	Daily closing prices in Euro (June 2020)												
	Number	10.6	11.6	12.6	15.6	16.6	17.6	18.6	19.6	22.6	23.6	24.6
German bank	50	8.71	8.10	8.22	8.29	8.44	8.42	8.22	8.26	8.34	8.51	8.12
BMW	12	58.81	55.00	56.17	56.32	57.11	57.34	57.08	56.97	57.68	58.49	56.40
Tesla	10	886.10	874.60	818.20	856.30	881.80	885.90	895.30	894.20	885.00	887.10	853.20
Allianz	30	185.40	175.68	177.66	177.80	183.02	181.90	181.00	180.40	179.64	182.98	178.36
Tui	100	5.50	4.88	5.08	5.12	5.29	5.25	5.06	4.95	4.72	4.69	4.30

Source: Finanzen.net, data downloaded from https://www.finanzen.net/ on 26.06.2020

Part III

From Few to All or from the Sample to the Population

We already know a lot about our observed enterprises. We know about their average behavior, we know about their variation, we have calculated correlations between variables and hence have information about relationships between variables. But does this knowledge about the observed enterprises apply to all newly founded enterprises? So far we have analyzed just a few enterprises. It could be that the results only apply to the observed ones. The central question is therefore whether our results can be applied to all newly founded enterprises. Technically speaking, the question is whether the sample results also hold for the population. In the next four chapters, we will address this question. How can we use knowledge about a few, the sample, to say something about all, the population? To answer this is the objective of the next four chapters. We look at our data again, clarify this crazy little thing called hypothesis -many people, even lecturers, often have a wrong idea about it-, look at how chance is distributed, and think around the corner with hypothesis testing. It's getting really exciting now.

Of Data and Truth

8

8.1 How do We Get our Data: Primary or Secondary Data?

It is now time to talk about our data again. Why do we have it and how do we get it? We have data to solve a problem or answer a question. Therefore, at the beginning there is always a problem or a question. We want to know something, for example:

- What is the average growth rate of start-ups?
- What is the average age of enterprise founders?
- Do more women or more men start businesses?
- Is there a relationship between marketing activities and the growth of newly founded companies?

or

- How much meat does the cheeseburger of McBurger contain?

So always the first thing we need to do is define the problem and, from that, the research question. Then we can think about what data we need to solve the problem and answer the research question. If the problem and the research question are not clearly defined, we usually run into two problems. Either we collect data we don't need, have extra work and create data graveyards, or we collect too little data and end up finding that the data we need hasn't been collected.

Once we have clearly defined the problem and the research question and are clear about what data we need, the next step is to get that data. There are two ways to do this. Either someone else has already collected the data and we can use it, or we have to collect the

© Springer-Verlag GmbH Germany, part of Springer Nature 2023
F. Kronthaler, *Statistics Applied With Excel*,
https://doi.org/10.1007/978-3-662-64319-8_8

Table 8.1 Data sources for secondary data

Institution	Internet address	Data level
Swiss Federal Statistical Office BfS	https://www.bfs.admin.ch/bfs/de/home.html	Data on Switzerland and the regions of Switzerland
Statistical Office of the European Union	https://ec.europa.eu/eurostat/de/home	Data on the European Union and its member states
OECD	https://data.oecd.org/	Mainly data on the member states of the Organisation for Development and Cooperation
World Bank	https://data.worldbank.org/	Data on almost all countries in the world
International Monetary Fund IMF	https://www.imf.org/en/Data	Mainly financial and economic data on almost every country in the world

data ourselves. In the first case we speak of secondary data, in the second case of primary data or primary data collection, we collect the data specifically for our research question.

Data collection is expensive and time-consuming, so whenever data is available, we should use it. There are a number of reputable data providers, not at least governmental and supranational institutions. At the government level, these are primarily the statistical offices of the states, in the EU the Statistical Office of the European Union, and at the supranational level the OECD, the World Bank and the International Monetary Fund (Table 8.1).

Let's take a quick look at the World Bank as an example. This gives an idea what a wealth of information is available. The World Bank provides online via the Internet address https://data.worldbank.org/ over 600 variables available in time series from 1960 onwards for almost all countries in the world. These range from poverty, climate change, agriculture and rural development to economic and social development (Fig. 8.1).

For example, we are interested (to stay close to our data set) in the start-up conditions of enterprises in different countries around the world. We want to know about the start-up conditions of Switzerland compared to other countries in the world. In the World Bank database we find the variable "Time required to start a business (days)". We download the data as an Excel file and now "only" have to analyze it (Table 8.2).

In addition to governmental and supranational institutions, there are a number of semi-governmental and semi-private institutions that offer data. Well-known examples include Transparency International, Freedom House and the World Economic Forum. Transparency International, for example, collects the Corruption Perception Index, which gives an indication of how large the problem of corruption is in individual countries of the world.

In any case, before we collect data ourselves, it is worth searching for data. If there is no data that we can use, we need to collect the data ourselves. We need to do primary data collection.

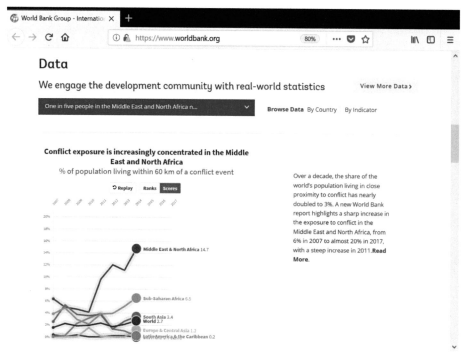

Fig. 8.1 The World Bank's data (14.04.2020)

8.2   The Random Sample: The Best Estimator for Our Population

If we are faced with the task of collecting data, the following aspects must first be clarified: (1) who is the population or about whom do we want to make a statement, (2) do we conduct a census or a random sampling, (3) if we conduct a random sampling, how do we achieve that the sample is representative for the population and (4) how large should the sample be.

First, we need to clarify who or what the population is. The population is all the people or objects we want to make a statement about. For example, this could be all the cheeseburgers that are produced or just the cheeseburgers that McBurger produces. In our example, the population consists of all newly founded enterprises.

Once we know what the population is, we have two options for collecting data. First, we can collect data on all persons or objects of the population, we do a census. Second, we can draw a subset out of all persons or objects, we do a sampling. Typically it is not possible to do a census for reasons of time and cost. In Switzerland alone, we have about 40'000 business start-ups per year. If we were to collect data on all 40'000 start-ups, this would firstly be very expensive and secondly very time-consuming—it would take at least a year to survey all foundations of a year. Moreover, even then we have not collected data

Table 8.2 WDI data on time required to start a business (Source: World Bank 2020, data downloaded from https://data.worldbank.org/indicator/IC.REG.DURS on 14/04/2020)

A	2010	2011	2012	2013	2014	2015	2016	2017	2018	2019
1 Data Source										
2 Last Updated Date										
3										
4 Country Name	2010	2011	2012	2013	2014	2015	2016	2017	2018	2019
38 Canada	5.5	5.5	5.5	5.5	5.5	1.5	1.5	1.5	1.5	
39 Central Europe and the Baltics	18.9090909	19.0909091	18.9090909	17.9090909	16.8181818	16.0909091	15.7272727	15.7272727	18	15.90
40 Switzerland	18	18	18	18	10	10	10	10	10	
41 Channel Islands										
42 Chile		36.5	9.5	7.5	7.5	7.5	7.5	7.5	7.5	6
43 China					32.3	29.3	29.3	26.9	22.9	8.5
44 Cote d'Ivoire	40	32	32	8	7	7	7	8	6	
45 Cameroon	20.5	16.5	16.5	16.5	16.5	16.5	16.5	16.5	13.5	
46 Congo Dem Rep	84.5	65.5	58.5	31.5	16.5	11.5	11.5	7	7	

B2: Time required to start a business (days)

A	B	C
1 INDICATOR_CODE INDICATOR_NAME	SOURCE_NOTE	
2 IC.REG.DURS	Time required to start a business (days)	Time required to start a business is the number of calendar days needed to complete the procedures to legally operate a business.
3		

on all start-ups, but only for one year. For all these reasons, we normally draw a sample out of the population, i.e. we take a subset and collect the required data for it. In our case, we have taken a sample of 100 companies from our population.

However, as mentioned at the beginning, we do not want to describe only the sample, but rather make a statement about the population. In order for this to be possible, the sample must behave in exactly the same way as the population with regard to the subject of the study, i.e. it must represent the population or be representative for the population. So the question arises is how do we achieve representativeness of the sample? The key criterion for this is that we draw the sample randomly out of the population. Technically speaking, each element of the population must have the same probability of being part of the sample. Let's imagine 20 people from which we want to select five. If we select these five at random, that is, if each person in the population of 20 has the same probability of being included in the sample, then we have satisfied this condition. If we make a mistake in the random selection, for example, if women have a higher probability of being selected, then the sample is biased and we can no longer make a statement about the population.

Freak Knowledge

We have several ways to select a representative sample from a population. The simplest is to draw a simple random sample. Each element of the population has the same probability of being selected into the sample. There is also stratified random sampling and cluster sampling procedure. In the former, the population is first divided into subgroups, e.g. by age or sex, and then a random sample is drawn from each subgroup. In the latter, we first divide the population into so called clusters and then randomly clusters are selected. We can also combine and then arrive at multistage procedures.

If each element of the population has the same probability of being part of the sample, then the requirement for a representative sample is met and the values obtained from the sample, e.g. the sample means, are the best estimators we have for the population. For us, this means, for example, that the average age of enterprise founders of 34.25 years, which we calculated for our 100 enterprises, is the best estimated value for the population. Technically speaking, the expected value of the sample is equal to the mean of the population:

$$E(\bar{x}) = \mu$$

where

$E(\bar{x})$ is the value we would expect when drawing a sample, and
μ is the mean of the population.

Let's think about this again for a moment. The value we get from our sample is the best value we can get. However, it is not the only value; it varies with the sample. To understand this, we again need to use the example of our 20 people. Let's imagine that we randomly draw five people from our population of 20 people. We have a total of 15'504 possibilities for this. The following figure illustrates four of these possibilities. We can randomly get the red marked people, the blue ones, the yellow ones or the green ones or we can mix the colors. If we think about it for a moment, we quickly notice that a large amount of possible combinations exist; as I said, there are 15'504 (Fig. 8.2).

Now let's think further and calculate the average body height for the different samples, e.g. the average body height of the red people, the yellow ones, and so on. It is fairly clear that we get different average body heights depending on the sample. Since there are 15'504 possible different samples, there are also 15'504 possible sample averages. However, we do not draw all possible samples, only one. From this one we say that it is the best estimator for the population. However, if we were to get a different sample, we would correspondingly get a different estimator. So we have a problem with that. We

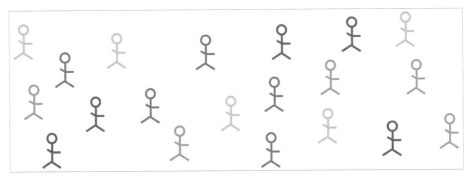

Fig. 8.2 Thought experiment on 5 out of 20

are not sure what the true value of the population is. The solution is that we integrate the uncertainty by not doing a point estimate, but by giving a range or interval in which the true value of the population is located with some confidence. To calculate the interval, we first need to be clear about something. If we draw all 15'504 samples and calculate all possible 15'504 sample means accordingly, we can in turn determine a "super" mean from the 15'504 sample means. All other possible sample means deviate around this mean. If all sample means are available, we can calculate the standard deviation of the sample means as well. However, we do not need all sample means at all, but only the values of one sample. We could show (but we will not do so here) that the standard deviation of the sample means can be estimated as follows (the little hat above the σ stands for estimated value):

$$\widehat{\sigma}_{\bar{x}} = \frac{s}{\sqrt{n}}$$

with

$\widehat{\sigma}_{\bar{x}}$ is the estimated standard deviation of the sample means, we also refer to this as the standard error of the mean,

$s = \sqrt{\dfrac{\Sigma(x_i - \bar{x})^2}{n-1}}$ is the standard deviation of the sample, and

n is is the number of observations.

From the sample mean and the standard error of the mean, we can calculate the interval in which the true value of the population probably lies. We speak of the confidence interval.

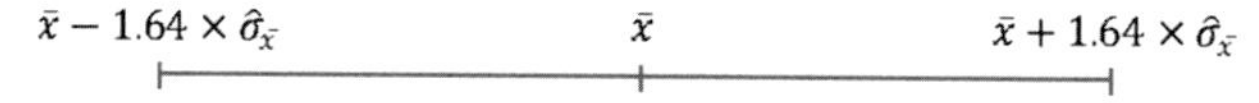

$$\bar{x} - 1.64 \times \hat{\sigma}_{\bar{x}} \qquad\qquad \bar{x} \qquad\qquad \bar{x} + 1.64 \times \hat{\sigma}_{\bar{x}}$$

Fig. 8.3 Graphical illustration of the 90% confidence interval of the mean

We usually apply three intervals, the 90%, 95% and 99% confidence interval:

$$CI_{90\%} = \left[\bar{x} - 1.64 \times \hat{\sigma}_{\bar{x}}; \bar{x} + 1.64 \times \hat{\sigma}_{\bar{x}}\right]$$

$$CI_{95\%} = \left[\bar{x} - 1.96 \times \hat{\sigma}_{\bar{x}}; \bar{x} + 1.96 \times \hat{\sigma}_{\bar{x}}\right]$$

$$CI_{99\%} = \left[\bar{x} - 2.58 \times \hat{\sigma}_{\bar{x}}; \bar{x} + 2.58 \times \hat{\sigma}_{\bar{x}}\right]$$

What is meant here is that the confidence interval has a 90%, 95%, 99% probability of covering the true value of the population. Or better still, for 90, 95, or 99 out of 100 samples, the value lies within this interval. Graphically, the confidence interval can be illustrated as shown in Fig. 8.3.

Let's calculate the 90%, 95%, and 99% confidence intervals for our example. We already know that the mean age of the enterprise founders, i.e. the sample mean, is 34.25 years. We have also already calculated the standard deviation with 7.62 years and the number of observations is 100. Thus, the confidence intervals can be displayed as follows:

$$CI_{90\%} = \left[34.25 - 1.64 \times \frac{7.62}{\sqrt{100}}; \bar{x} + 1.64 \times \frac{7.62}{\sqrt{100}}\right] = [33.00: 35.50]$$

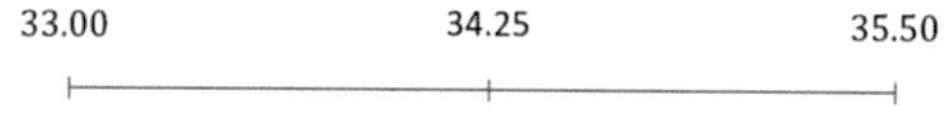

33.00   34.25   35.50

$$CI_{95\%} = \left[34.25 - 1.96 \times \frac{7.62}{\sqrt{100}}; \bar{x} + 1.96 \times \frac{7.62}{\sqrt{100}}\right] = [32.76: 35.74]$$

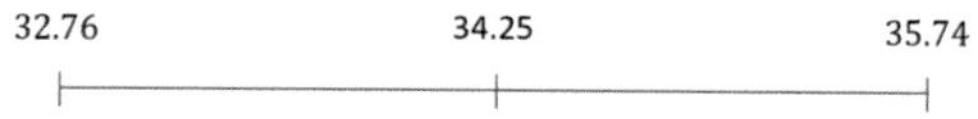

32.76   34.25   35.74

$$CI_{99\%} = \left[34.25 - 2.58 \times \frac{7.62}{\sqrt{100}}; \bar{x} + 2.58 \times \frac{7.62}{\sqrt{100}}\right] = [32.28: 36.22]$$

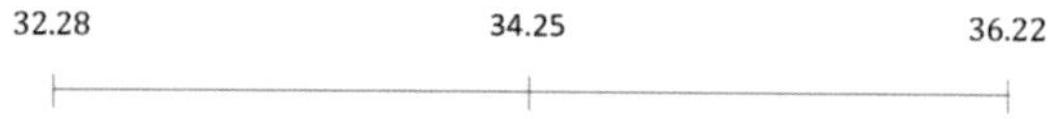

32.28   34.25   36.22

We can thus make the following statement. The calculated confidence interval of 33.00 to 35.50 covers the true value of the population with a probability of 90%. If we want to increase the certainty, the confidence interval becomes wider. If we want to be 99% certain that the confidence interval overlaps the true value, then the confidence interval ranges from 32.28 to 36.22.

We can also use this information to say something about the sample size, that is, how many people or objects we should observe. For the confidence intervals described above, we can determine the width. It is two times the distance from the mean to the upper or lower interval endpoint. For the 95% confidence interval, the confidence interval width is for example

$$CIW_{95\%} = 2 \times 1.96 \times \widehat{\sigma}_{\bar{x}} = 2 \times 1.96 \times \frac{s}{\sqrt{n}}$$

Let's solve for n, we get the formula that we can use to work out the sample size we need:

$$n = \frac{2^2 \times 1.96^2 \times s^2}{CIW_{95\%}{}^2}$$

where

n	is the sample size,
s	is the standard deviation of the sample, and
$CIW_{95\%}$	is the confidence interval width of the 95% confidence interval.

We observe the following. First, our sample size depends on the confidence interval width. Second, it depends on the standard deviation of the sample, better the standard deviation in the population. Originally, we had used the standard deviation of the sample as an estimate for the standard deviation of the population. We need to determine the latter from experience or from prior research. Third, the sample size depends on how large we choose our confidence level. At 90%, instead of 1.96, we take 1.64; at 99%, we take 2.58. Fourth, and this is the amazing thing, the sample size does not depend on the size of the population. Unless the population is very small, the size of the population can be neglected. If the population is very small, we will usually conduct a census and the sampling problem does not arise.

Let's apply the formula to our example. We want to estimate the average age of enterprise founders at the 95% confidence level and the confidence interval width should not be larger than 3 years. We know from prior research that the standard deviation in the population is about 8 years. Putting in the values, we get the required sample size:

$$n = \frac{2^2 \times 1.96^2 \times s^2}{CIW_{95\%}{}^2} = \frac{2^2 \times 1.96^2 \times 8^2}{3^2} = 109.3$$

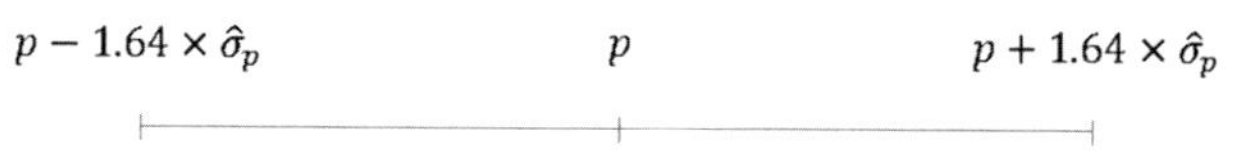

$$p - 1.64 \times \hat{\sigma}_p \qquad\qquad p \qquad\qquad p + 1.64 \times \hat{\sigma}_p$$

Fig. 8.4 Graphical illustration of the 90 % confidence interval for proportion values

That is, in order to comply with the given framework, we need to draw a random sample of about 109 observations. Then we obtain the true value in the population with a 95% certainty in an interval of 3 years.

We can easily apply the considerations just made for the mean value to proportion values. There are only a few small changes. For a representative sample, the expected value for the proportion value of the sample is equal to the proportion value of the population:

$$E\,(p) = \pi$$

with

$E\,(p)$ is the proportion value we expect when drawing the sample, and
π is the proportion value of the population.

The standard deviation for the sample proportion values respectively the standard error of the proportion value is estimated as follows:

$$\hat{\sigma}_p = \sqrt{\frac{p\,(1 - p)}{n}}$$

with

$\hat{\sigma}_p$ is the standard error of the proportion value and
p is the proportion value of the sample.

The 90% confidence interval is calculated from the calculated sample proportion value p and its standard error (Fig. 8.4). If we want to calculate the 95 % confidence interval instead of the 90% confidence interval, we take 1.96 again instead of 1.64, or 2.58 for the 99% confidence interval.

Let's illustrate this with another example. From our data set, we can easily see that 35% of the firms were founded by a woman. In decimal notation this means $p = 0.35$.

If we put the proportion value into our formula for the standard error, we get a standard error of rounded 4.8 percentage points:

$$\hat{\sigma}_p = \sqrt{\frac{p\,(1 - p)}{n}} = \sqrt{\frac{0.35\,(1 - 0.35)}{100}} = 0.048$$

The 90% confidence interval thus has a range from 27.1% to 42.9%.

$$CI_{90\%} = [0.35 - 1.64 \times 0.048; 0.35 + 1.64 \times 0.048] = [0.271; 0.429]$$

27.1%   35%   42.9%

Hence, with 90% confidence, the true proportion of female founders in start-ups is between 27.1% and 42.9%. If we increase the confidence level to 95% or to 99%, the confidence interval becomes wider. For the 95% confidence interval the range is between 25.6% and 44.4%, for the 99% confidence interval 22.6% and 47.4%.

If we look at the confidence intervals, we may have the impression, due to their seize, that we are relatively imprecise with respect to the proportion value in the population. If we want a more precise statement, we have to increase the size of the sample.

How large the sample needs to be depends again among others on the confidence interval width we want to allow. So we use the formula above again, with the small difference that we insert the standard error of the proportion value. In the following, the formula describes the sample size at a confidence level of 90%. If we want to achieve a confidence of 95% or 99% instead, we replace the number 1.64 with 1.96 or 2.58 respectively.

$$CIW_{90\%} = 2 \times 1.64 \times \hat{\sigma}_p = 2 \times 1.64 \times \sqrt{\frac{p\,(1-p)}{n}}$$

Let's solve the formula for n, we obtain the formula for the sample size at a confidence interval of 90%.

$$n = \frac{2^2 \times 1.64^2 \times p(1-p)}{CIW_{90\%}^2}$$

Let's illustrate this again with an example. We assume that the confidence interval width of the 90% confidence interval should not exceed six percentage points. To fill in the formula completely the roportion value p is missing. Before we draw the sample, we do not know the proportion value. Hence, we need to determine it using reasoning and previous studies. For our example, let's assume that the proportion of female founders is around 30%. If we now insert the values, we obtain a sample size of

$$n = \frac{2^2 \times 1.64^2 \times p(1-p)}{CIW_{90\%}^2} = \frac{2^2 \times 1.64^2 \times 0.3(1-0.3)}{0.06^2} = 627.6.$$

Thus, to be able to say with 90% confidence that the true proportion value of the population lies within a range of 6 percentage points, we need to draw a sample of $n = 627.6$.

We have seen, the size of the sample essentially depends on how precise we want to be in our statement. This has to be determined in advance of the sampling.

8.3   Of Truth: Validity and Reliability

Drawing a good sample is only one side of the coin in learning about the population. Equally important is that we collect what we want to know and that we do so reliable. In technical terms, we talk about validity and reliability of data.

Validity, in simple terms, means that the data actually measure what is the matter of the study. Sometimes this is easy, sometimes not. For example, if we are interested in the profit of young companies, then we can collect the profit in monetary units and the matter is relatively clear. If we want to know about the difference in wealth of countries, then the matter is no longer so simple. People often use the concept of gross domestic product per capita to measure the wealth of a country. In simple terms, gross domestic product per capita measures the goods and services that are produced and available for consumption in a country per person. Therefore, it is often taken as a proxy for the wealth of a country. Proxy is the technical term for a variable that gives approximate information about the issue in which we are interested. However, many people doubt that GDP per capita is a good variable to tell about the wealth of people. Internationally, for example, the Human Development Indicator is also used. This indicator is a composite measure that includes not only GDP but also, for example, life expectancy in a country, medical care and education. The government of Bhutan uses the "gross national happiness". While it may seem odd, happiness can perhaps also be a proxy for wealth. It is relatively easy to see that depending on the indicator, different results will emerge when we say something about the wealth of countries. Let's imagine that we have to measure happiness. The task would then first to define what happiness is. Then we need to find a proxy that actually measures happiness. That seems complicated, does it not?

How can we ensure to find an appropriate measure that includes what we want to know. The easiest way is to go to an expert and ask what measure can be used. So if we want to know something about corporate innovation activity and we are not innovation experts ourselves, then we go to a person who has been studying corporate innovation for some time. If no one is available, then we have no choice but to read the literature to find out how innovation can be approximately measured in a business context. There will certainly not be only one way. By the way reading the literature is always recommended, you reduce your dependency on others or you can better understand and evaluate what they say.

Once we have ensured that we have a valid variable or construct to answer our question, then we need to ensure that we measure it reliable, we need to ensure the reliability of the data. Reliability, in simple terms, means that we measure the true value and that when we measure it several times, we will always get the same result. As a simple example, let's say that we want to measure the intelligence quotients of enterprise founders. We must first realize that any measurement is subject to error. So let's say we pick a founder of an enterprise and survey the intelligence quotient for him. Meanwhile, we also have a construct, a written test, that allows us to measure it. Let's further assume that the true intelligence quotient of this founder is 115 points. We do not know this value, we are trying to measure it. Our test result indicates that the intelligence quotient is 108. We then made an error of seven points. This error can occur because the founder was tired, because he was distracted, because the room temperature was too high, because the questions were too complicated, etc. It follows that we have to create conditions that ensure that the error is as small as possible. The smaller the error, the more reliable the measurement. Establishing reliability requires that we ensure that we are collecting the true value and that we do it with an error as small as possible.

Validity and reliability are closely related. If we do not take either into account, our study is worthless. We would not be able to answer our research question. Let's imagine that we take care of validity, but do not consider reliability. Then we are measuring the right thing, but we do not measure it correctly. If we take reliability into account, but not the validity of the data, we measure correctly, but the wrong thing. The data are then worthless with respect to our research question. The most we could then do is answer a different question that fits the data. With regard to our analysis, this means that we must first ensure validity and then reliability.

8.4   Checkpoints

- *Before we collect and analyse data the research problem and the research question have to be clearly defined.*
- *Before primary data collection is carried out we should check whether the required data are available by a secondary source.*
- *The population are all persons or objects about we want to make a statement.*
- *We achieve representativeness of the sample by randomly drawing the elements or objects out of the population.*
- *The sample values of a representative sample are the best estimators for the population.*
- *The confidence interval makes the point estimate more precise by including uncertainty.*
- *The sample size is calculated from the width of the confidence interval, the standard deviation in the population and the defined confidence level.*
- *Validity of a variable means that the variable is a good proxy for the object of interest.*
- *Reliability means that we measure the object of interest with an measurement error as small as possible.*

8.5   Applications

8.1 What steps do we need to follow when collecting data?

8.2 Search data on the population of countries on the Internet. How many inhabitants do China, India, Switzerland, Germany and the USA currently have?

8.3 Explain what the population is, what a sample is, and what condition must be satisfied so that we can infer from the sample to the population.

8.4 What is the average growth rate of firms and what is the 90%, 95% and 99% confidence interval, data set data_growth.xlsx?

8.5 Calculate the sample size needed if we want to estimate the average growth rate of firms with a 95% confidence level and a confidence interval width not wider than 2 percentage points. From preliminary research we know that the standard deviation in the population is about 5 percentage points.

8.6 What percentage share of the enterprise in our data set data_growth.xlsx are industrial enterprises and what is the 90%, 95% and 99% confidence interval?

8.7 What would the sample size need to be if we want to estimate the share of industrial enterprises in the population with a 90% confidence level and a confidence interval width of 8 percentage points. From preliminary considerations, we know that the share of industry enterprise is around 25%.

8.8 What is validity and reliability?

8.9 Which of the two statements is correct: 1) reliability is possible without validity or 2) validity is possible without reliability. Explain briefly.

8.10 Locate the article by Costin, F., Greenough, W.T. & R.J. Menges (1971): Student Ratings of College Teaching: Reliability, Validity, and Usefulness, Review of Educational Research 41(5), pp. 511–535 and discuss how this article discusses the concepts of validity and reliability.

9.1   The Little, Big Thing of the (Research) Hypothesis

Hypothesis is a term that is often used in statistics and research, but sometimes it seems that it is still not entirely clear to people. Even lecturers sometimes have a wrong idea of what a hypothesis is. A hypothesis is basically something very simple. Plainly spoken, it is just a translation of the research question into a testable statement. At the same time, however, a hypothesis is also something quite grand. With a hypothesis, we generate statistically validated knowledge.

Let's consider this with an example. We'll stress the burger example again. Our question is as follows: How much meat does a burger from McBurger contain? With the help of the previous chapter we can answer this question with a point and an interval estimation. If we have previous knowledge, the question can be made more precise and formulated as a statement. Let's assume that McBurger states that a burger contains on average 100 grams of meat. With this information, we can translate our question into a statement. The statement or hypothesis which can be tested is then as follows: A burger from McBurger contains on average 100 g meat. A hypothesis is created by formulating the question more precise.

- *We come from the question:*
- *How much meat does a burger from McBurger contain?*
- *via previous knowledge to ...*
- *the hypothesis: A burger from McBurger contains on average 100 g meat.*

The advantage of the hypothesis over the question is that it is more precise and can therefore be tested. What does this mean in concrete terms? In the previous chapter, we did a point estimate and an interval estimate to answer the research question. After collecting

© Springer-Verlag GmbH Germany, part of Springer Nature 2023   125
F. Kronthaler, *Statistics Applied With Excel*,
https://doi.org/10.1007/978-3-662-64319-8_9

the data, we might have found that the sample mean is 98 g and the 95% confidence interval ranges from 96 g to 100 g.

Would that have made us reject McBurger's statement? We don't know really. We did not test the statement, but generated knowledge about how much meat a burger from McBurger contains. With the help of a hypothesis, we can test the statement. We do this (as we will see in the following chapters) using knowledge generated from the sample. If we find that the statement is not true, we can confront McBurger and ask for more or less meat, depending on the result.

How do we come, however, from the question to the hypothesis? Hypotheses do not arise from the gut. We form them through existing knowledge in the form of studies and theories. Let's look at this using our start-ups as an example. The question we are interested in is as follows. What is the average age of enterprise founders? We search for literature on enterprise start-ups and find studies that have been done in the past. We also come across theories on start-ups, which state that founders have gained sufficient professional experience before starting the business. They are typically not at the beginning but in the middle of their professional career. Based on this pre-existing knowledge, we formulate our hypothesis. This could be as follows: Enterprise founders are on average 40 years old.

- *We come from the question:*
- *How old are enterprise founders on average?*
- *via existing studies and theories to ...*
- *the hypothesis: Enterprise founders are on average 40 years old.*

Let's recap: A hypothesis is a sound testable statement formulated with the help of theory and previous knowledge. In the research process, we can also refer to the hypothesis as the research hypothesis.

9.2   The Null Hypothesis H_0 and the Alternative Hypothesis H_A

We now know what a hypothesis is and how we generate the research hypothesis. However, this is not enough, in the research process, we divide the research hypothesis into two different types, the null hypothesis and the alternative hypothesis. The null hypothesis is always the testable statement, the alternative hypothesis is the opposite of the null hypothesis. Let's look at our burger example again to clarify.
Research hypothesis: A burger from McBurger contains on average 100 g meat.
If we divide this statement into the null hypothesis and the alternative hypothesis, we get the following two statements.

H_0:   *A burger from McBurger contains on average 100 g meat.*

H_A: *A burger from McBurger contains on average more or less than 100 g meat.*

For the null hypothesis we write H_0 and for the alternative hypothesis H_A. Sometimes the alternative hypothesis is also denoted as H_1.

For our enterprise founders, the research hypothesis was as follows.

Research hypothesis: Enterprise founders are on average 40 years old.

If we divide this statement again into the null hypothesis and the alternative hypothesis, then we get the two following statements.

H_0:   *Enterprise founders are on average 40 years old.*

H_A: *Enterprise founders are on average younger or older than 40 years.*

We need the distinction between null hypothesis and alternative hypothesis for two reasons, first for precision. We need to know what happens when we test the null hypothesis and reject it. For the second example, let's test H_0 and assume that we reject it. We can then make the statement that enterprise founders are younger or older than 40. If we do not reject H_0 we remain with the statement that company founders are on average 40 years old.

Second, it is typically not possible to test the hypothesis formed from previous knowledge and theories directly. We then have to go via the null hypothesis. To illustrate this, we will extend the example of the average age of enterprise founders. We are interested in the following question. Are male and female founders of the same age when they start an enterprise? We search the literature, find studies and theories and form our research hypothesis accordingly. This then is, for example, as follows.

Research hypothesis: Male and female founders are not of the same age when they start an enterprise.

Can we test this hypothesis? What does not equal old mean for the population? Is it enough to find a difference of 0.5 years in the sample or does the difference have to be 5 years so that we come to this conclusion from the sample? As you can see, the statement is imprecise and thus cannot be tested, not the same age is not precisely defined. What can be tested, however, is the statement that male and female founders are of the same age when they found their enterprise. Equal age is clearly defined, the difference is zero years. Our research hypothesis that male and female founders are of different ages when they start a business is not testable, but the opposite is testable. In this case, we make the opposite of the research hypothesis to the null hypothesis. If we reject the null hypothesis based on the sample results, we proceed with the alternative hypothesis, which is equal to the research hypothesis.

H_0:   *Male and female founders are of the same age when they start an enterprise.*

H_A: *Male and female founders are not of the same age when they start an enterprise.*

Next, let's use our enterprise example to explain the principle using an assumed correlation between two variables. For this, we use the following research question. Is there a correlation between the resources young enterprises spend on marketing and the resources they spend on innovation? We review the literature and formulate the following hypothesis based on the results.

Research hypothesis: There is a relationship between the resources young companies spend on marketing and the resources they spend on innovation.

This hypothesis implies a correlation between the two variables marketing and innovation. But what exactly does a correlation mean for the population? Does the correlation coefficient found in the sample have to be 0.2, 0.5 or 0.7? The statement is imprecise. What is precise, however, is the statement that there is no correlation between marketing and innovation. No correlation means the correlation coefficient is exactly zero. The null hypothesis is thus again the opposite of the research hypothesis.

H_0:   There is no relationship between the resources young companies spend on market- ing and the resources they spend on innovation.

H_A: There is a relationship between the resources young companies spend on marketing and the resources they spend on innovation.

9.3   Hypotheses, Directional or Non-directional?

The null hypothesis and the alternative hypothesis can be formulated non-directional or directional. Whether we formulate a null hypothesis and an alternative hypothesis directional or non-directional depends on how much we know. If we know a lot, we can formulate the null hypothesis and the alternative hypothesis typically directional, if we know less, we are content to formulate them non-directional. The best way to show this is again to use our two examples just given. We have so far formulated the hypotheses in an undirected way in both examples. In the first example, we stated that there is a difference respectively no difference in age between male and female enterprise founders. In the second case, we formulated no or a relationship between marketing and innovation. We did not postulate a direction in either case, that is, we did not say whether men or women were older or whether the correlation is positive or negative. If we add this information, we come to a directional null hypothesis and alternative hypothesis. So to formulate hypotheses directional, we need more information or more knowledge. Let's imagine that the literature review has led to a higher level of information in both cases, e.g. we reasonably assume that women are older than men when they start an enterprise or that there is a negative correlation between the two variables marketing and innovation. Then we are able to formulate our research hypothesis, null hypothesis and alternative hypothesis as follows.

Research hypothesis: Female founders are older than male founders when starting an enterprise.

This results in the following null and alternative hypothesis, where the research hypothesis becomes the alternative hypothesis.

H_0: *Female founders are younger or the same age as male founders when starting an enterprise.*
H_A: *Female founders are older than male founders when starting an enterprise.*

In the second example, the research hypothesis is as follows.

Research Hypothesis: There is a negative relationship between the resources young companies spend on marketing and the resources they spend on innovation.
Again, the research hypothesis becomes the alternative hypothesis.

H_0: *There is no or a positive relationship between the resources young companies spend on marketing and the resources they spend on innovation.*
H_A: *There is a negative relationship between the resources young companies spend on marketing and the resources they spend on innovation.*

In summary, in non-directional hypotheses, the direction is not specified, while in directional hypotheses, the direction is also specified. The null hypothesis includes not only the opposite of the alternative hypothesis, but also the equality sign.

9.4 How to Formulate a Good Hypothesis?

Now we should still briefly discuss what makes a good hypothesis. Basically, a good hypothesis should meet the following criteria. First, it should be an extension of the research question. Hypotheses should always be closely related to the research problem respectively the research question. With their help, we want to solve the problem or find answers to our question. Secondly, hypotheses have to be formulated as a statement, not as a question. Questions are not testable, but clear, precise statements are. If we formulate the statement that a burger contains 100 g of meat, that statement can be tested. Third, a good hypothesis reflects the existing literature and theory. Almost always, we are not the only ones who are concerned with the research problem. Before, typically many others have addressed similar questions. The resulting knowledge helps to answer our question and to formulate our hypothesis as precisely as possible. A good hypothesis always has a strong relation to the existing literature. Fourth, a hypothesis must be brief and meaningful. If the hypothesis contains no statement, then nothing can be tested. The longer and thus usually more complicated a hypothesis is, the more difficult is it to formulate the hypothesis precisely and unambiguously. Fifth and finally, the hypothesis must be testable. Let's imagine that we have formulated the following research hypothesis. Women are better enterprise founders than men. In this hypothesis, it is not clear what is meant by better. Do women have the better business model? Do women have a better process of starting an enterprise? Do enterprises founded by women grow faster? etc. We don't know what to test. In contrast, the following formulation of the research hypothesis is much more precise and therefore testable. Firms founded by women achieve an higher turnover growth in the first five years than firms founded by men. Here it is clear that we need to compare turnover growth.

9.5   Checkpoints

- *A hypothesis is a statement about an issue based on theory or experience.*
- *In the research process we distinguish between null hypothesis H_0 and alternative hypothesis H_A. The null hypothesis is the testable statement, the alternative hypothesis is the opposite of the null hypothesis.*
- *We can formulate our hypotheses non-directional and directional. Formulating a directional hypothesis requires typically a higher level of knowledge than formulating a non-directional hypothesis.*
- *A good hypothesis meets the following criteria: It is an extension of the research question, it is formulated as a statement, it reflects the existing literature, and it is precise and testable.*

9.6   Applications

9.1 Formulate the non-directional null and alternative hypotheses for the following research questions.

- Is there a relationship between the turnover growth of a start-up firm and the proportion of turnover spent on innovation?
- Is there a difference in industry experience between male and female founders, which they possess when starting an enterprise?
- Is there a correlation between smoking and life expectancy?

9.2 Formulate the directional null and alternative hypothesis for each of the research questions above.

9.3 Go to the library and find three journal articles (which contain data) from your field of studies. For each article, answer the following questions: What is the full citation? What is the research question of the article? What is the research hypothesis? Is it explicitly stated? What is the null hypothesis? Is it explicitly stated? In addition, for those articles in which the hypothesis is not explicitly stated, formulate the research hypothesis.

9.4 Evaluate the hypotheses you found in Task 3 in light of the five criteria that a good hypothesis should meet.

9.5 Why does the null hypothesis usually assume no relationship between variables or no difference between groups?

9.6 If interested, read the story "Die Waage der Baleks" by Heinrich Böll and formulate a hypothesis about the story with in mind the weighing machine. Think about how the boy tests his presumption. Unfortunately the book is only available in German.

Normal Distribution and Other Test Distributions 10

We now know what a hypothesis is and that we use them to answer research questions or test statements. However, we do not yet know how testing works. To understand this, we must first realize that chance is not completely random. We will see that chance can be normally distributed, but it can also be distributed differently. That means very specifically, we can make a statement about how random an random event occurs. Interesting, isn't it? In the following, we will deal with probability distributions. We will learn about the normal distribution and about other test distributions. It remains exciting.

10.1 The Normal Distribution

The normal distribution is interesting for two reasons. First, in reality many variables are at least approximately normally distributed. Second, the normal distribution gives us an intuitive approach to probability distributions. Once we are used to the normal distribution, dealing with other test distributions is no longer a problem.

Let's first describe the normal distribution. The normal distribution is bell-shaped, with a maximum. It is symmetric around the maximum. Its tails converge asymptotically to the x-axis (Fig. 10.1).

The maximum of the normal distribution is in the middle of the curve. Symmetric means that we can fold the normal distribution in its middle and then both sides lie exactly on top of each other. For both reasons, the arithmetic mean, median, and mode are equal. The mode lies in the middle because it is the most frequent value. The median lies in the middle because the distribution is symmetrical and the median divides the distribution exactly in two halfs. Finally, the mean is in the middle because we have exactly the same amount of values to the left and to right, meaning we have exactly the same amount of deviations in both directions. Asymptotic means that the tails get closer and closer to the

© Springer-Verlag GmbH Germany, part of Springer Nature 2023
F. Kronthaler, *Statistics Applied With Excel*,
https://doi.org/10.1007/978-3-662-64319-8_10

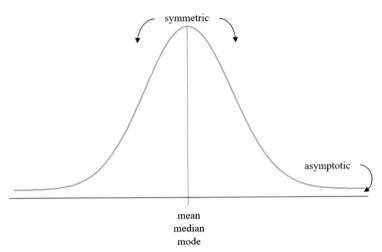

Fig. 10.1 The normal distribution

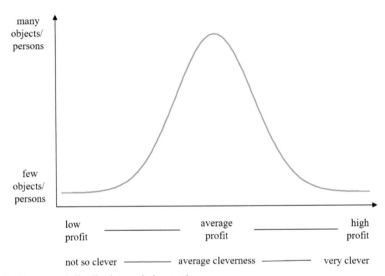

Fig. 10.2 The normal distribution and observations

x-axis, but never reach it. The area under the normal distribution curve indicates how likely it is that a particular value of a normally distributed variable will occur within an interval.

Let's illustrate this with two variables for which we can easily assume that they are normally distributed: the intelligence of individuals and the profit of enterprises in an industry. Let's have a look at Fig. 10.2 and start with the intelligence quotient. We keep in mind that the area under the normal distribution curve reflects the probability that a given value occurs within an interval.

In the figure, the number of people is plotted on the y-axis and the intelligence quotient is plotted on the x-axis, where we have people with a low intelligence quotient on the left, and people with a high intelligence quotient on the right. The people with an intermediate intelligence quotient are in the middle. We know from experience that very smart people and very dumb people are very rare. The curve reflects that. The areas far to the left and far to the right under the normal distribution curve are very small, and the curve is close to the x-axis. People who are of average intelligence, on the other hand, are very common. The central area under the normal distribution curve is very large. It is the same with enterprise profits. We have many companies that make average profits, few that perform exceptionally well, and also few with exceptionally poor performances. The latter ones will disappear from the market soon or later. Accordingly, the areas to the right and left under the normal distribution curve are small, and the central area in the middle is large.

Let's look at this in more detail. We can easily divide the normal distribution into intervals. In the figure below, we have done this for a normal distribution with mean 100 and standard deviation of 20. 100 is the mean, 120 and 80 are 1 standard deviation (1 × 20) away from the mean, 140 and 60 are 2 standard deviations (2 × 20) away from the mean, and 160 and 40 are 3 standard deviations (3 × 20) away from the mean (Fig. 10.3).

If we look at the figure, we see that the two central areas between 80 and 100 and between 100 and 120 each have an area of 34.13% of the total area. Jointly together the area is 68.26%. In other words, the area from minus one standard deviation to one standard deviation is 68.26% of the total area. What does this mean specifically? It means that values that deviate no more than one standard deviation from the mean have a chance to occur of 68.26%.

Looking at our intelligence quotient example, we can make the statement that 68.26% of people have an intelligence quotient between 80 and 120 points. If we apply this to our sampling and pick any person out of all the people, then there is a 68.26% chance to find someone whose intelligence quotient falls within this interval. We can continue in exactly the same way. The probability of finding a person whose intelligence quotient is

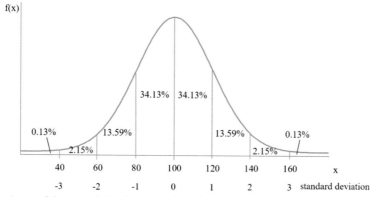

Fig. 10.3 Areas of the normal distribution and standard deviation

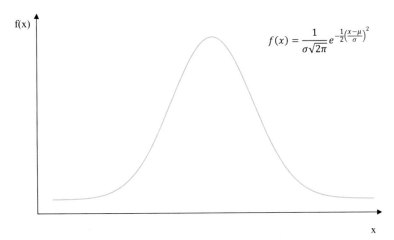

Fig. 10.4 The function of the normal distribution

in the interval between 120 and 140 points or between one and two standard deviations is 13.59%, for an intelligence quotient between 140 and 160 or between two and three standard deviations it is 2.15% and for more than 160 points or more than three standard deviations from the mean it is 0.13%. Of course, we can do the same when we go to the left of the mean.

So far we used a special normal distribution, the normal distribution with mean 100 and standard deviation 20. One small problem with this is that we do not have only one normal distribution, we have a variety of normal distributions. So we need to go further and look at the normal distribution again in more detail.

In Fig. 10.4 the normal distribution curve function is displayed.

The function is as follows:

$$ f(x) = \frac{1}{\sigma\sqrt{2\pi}} e^{-\frac{1}{2}\left(\frac{x-\mu}{\sigma}\right)^2} $$

where

$f(x)$ is the function value at the point x, i.e. the value on the y-axis at the position of the respective x-value,

μ is the mean of the normal distribution,

σ is the standard deviation of the normal distribution,

π is 3.141...,

e is Euler's number 2.718....

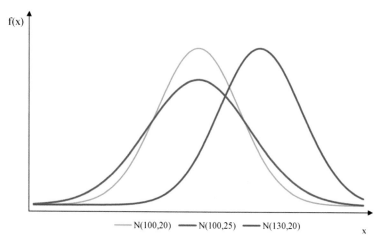

Fig. 10.5 Comparison of the normal distributions $N(100, 20)$, $N(100, 25)$ and $N(130, 20)$

From this we can conclude that the function values and hence the visual nature of the normal distribution depend on only two parameters, the mean and the standard deviation. With these two values, we can uniquely define a normal distribution. Therefore, a normally distributed variable can be characterized as follows:

$$X \sim N\,(\mu, \sigma)$$

The variable X e.g. our intelligence quotient or our enterprise profits, is normally distributed (the tilde stands for distributed like) with the mean value μ and the standard deviation σ.

When the mean or standard deviation change, the shape of the normal distribution also changes.

Specifically, increasing the mean causes the normal distribution curve to shift to the right, and increasing the standard deviation causes it to widen. This can be easily seen in Fig. 10.5. The normal distribution curve $N(130, 20)$ is further to the right than $N(100, 20)$, and the normal distribution curve $N(100, 25)$ is wider than $N(100, 20)$ and $N(130, 20)$. Since both the mean and the standard deviation can theoretically take on any possible value, there is not just one normal distribution, but infinitely many.

10.2   The z-Value and the Standard Normal Distribution

We have discussed that we do not have only one normal distribution, but that there are many normal distribution whose shape and location depend on the mean and standard deviation. However, we can standardize each of these normal distributions with a simple transformation so that they have a mean of 0 and a standard deviation of 1. The

Table 10.1 Standardization of the normal distribution $N(100, 30)$

x-value	Transformation	z-value
10	$z = \frac{10-100}{30}$	−3
25	$z = \frac{25-100}{30}$	−2.5
40	$z = \frac{40-100}{30}$	−2
55	$z = \frac{55-100}{30}$	−1.5
70	$z = \frac{70-100}{30}$	−1
85	$z = \frac{85-100}{30}$	−0.5
100	$z = \frac{100-100}{30}$	0
115	$z = \frac{115-100}{30}$	0.5
130	$z = \frac{130-100}{30}$	1
145	$z = \frac{145-100}{30}$	1.5
160	$z = \frac{160-100}{30}$	2
175	$z = \frac{175-100}{30}$	2.5
190	$z = \frac{190-100}{30}$	3

transformation rule is as follows:

$$z = \frac{x - \mu}{\sigma}$$

with

z is the standardized value (z-value for short),
x is the value of the distribution to be standardized,
μ is the mean value of the distribution,
σ is the standard deviation of the distribution.

In Table 10.1, we apply the transformation rule to the normal distribution $N(100, 30)$ with mean of 100 and standard deviation of 30. Instead of this one, we could use any other normal distribution. We see that using the z-transformation rule, the mean of 100 become 0 and the new standard deviation is 1.

With the z-transform, we transformed the normal distribution $N(100, 30)$ into the normal distribution $N(0, 1)$. We also call the latter the standard normal distribution, with the z-values are now plotted on the x-axis instead of the x-values. However, the interpretation remains the same. In the interval from − 1 standard deviation (z-value of − 1) to + 1 standard deviation (z-value of + 1) lie 68.26% of the possible values. That is, the probability that we find a standardized z-value in the interval from − 1 to + 1 is

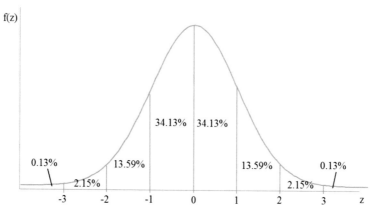

Fig. 10.6 The standard normal distribution $N(0, 1)$

68.26%, that we get a standardized z-value greater than 3 is 0.13%, and so on. Compare Fig. 10.6.

But what does the z-value really mean and why do we need it? The z-value always represents an x-value of any normal distribution. With its help we can determine for any normal distribution how likely the occurrence of a certain x-value and larger is. An example will illustrate this. Assume the intelligence quotient is normally distributed with a mean of 100 and a standard deviation of 10. We now find a person with an intelligence quotient of 130. The z-value representing 130 is

$$z = \frac{130 - 100}{10} = 3.0.$$

Using this z-value and the standard normal distribution, we can make a statement about how likely it is to find a person with a intelligence quotient of 130 and greater. The probability is 0.13% (compare Fig. 10.6). We can also ask ourselves what is the probability of finding a person with a intelligence quotient between 80 and 90 points. The corresponding z-values are

$$z = \frac{80 - 100}{10} = -2.0 \text{ and } z = \frac{90 - 100}{10} = -1.0,$$

i.e. the probability of finding such a person is 13.59%.

Any given area of the standard normal distribution automatically has a given z-value associated with it. Thus, using the z-value, we can determine the area under the standard normal distribution. As can be seen in Fig. 10.7, for example, a z-value of 1.18 divides the standard normal distribution into the areas of 11.9% to the right and 88.1% to the left.

As the z-value becomes larger, the area to the right decreases and the area to the left increases. Likewise, the reverse is true.

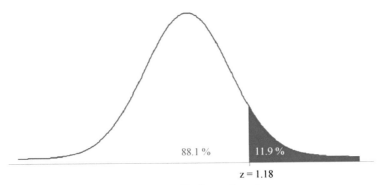

88.1 %   11.9 %

$z = 1.18$

Fig. 10.7 Area under the standard normal distribution and z-value

Since calculating the area is basically very time consuming, thoughtful minds have tabulated the standard normal distribution for us, its areas, and its z-values (see Appendix B). As we will see in the following chapter, the z-value is important to test hypothesis. With its help, we can state how likely or unlikely the occurrence of a particular x-value is and (as we will see in the next chapter) reject or do not reject hypotheses.

10.3   Normal Distribution, t-Distribution, χ^2-Distribution and (or) F-Distribution

Variables are not only normally distributed, but can also be distributed in other ways. At this point, therefore, we want to introduce three more important distributions that we need for testing: the t-distribution, the χ^2-distribution (pronounced chi-squared), and the F-distribution. Basically, we use these distributions just like the standard normal distribution when testing. Again, the distributions are tabulated (see Appendices 3–5).

The t-distribution is symmetrical like the standard normal distribution with a maximum, a mean of zero and asymptotic towards the x-axis. However, it is slightly wider and flatter, how wide and how flat depends on the number of degrees of freedom (df). If we have few degrees of freedom, the t-distribution is more flat and more wide, if we have more degrees of freedom, it becomes narrower and more peaked. From about 30 degrees of freedom on the t-distribution is equal to the standard normal distribution (compare Fig. 10.8).

We will discuss the meaning of the degrees of freedom later when we need the t-distribution in testing. It is important to say here that we need the t-distribution in practical work, for example, when we are dealing with small samples.

In addition to the t-distribution, the χ^2-distribution is another important test distribution. We need it especially when testing nominal variables. As with the t-distribution, there is not just one χ^2-distribution, but many χ^2-distributions which depend on the number of degrees of freedom. We will come to the degrees of freedom when we use the χ^2-

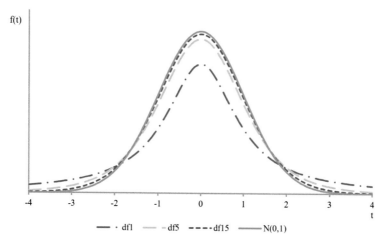

Fig. 10.8 The *t*-distribution compared to the standard normal distribution

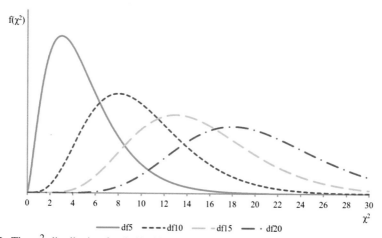

Fig. 10.9 The χ^2-distribution for 5, 10, 15 and 20 degrees of freedom

distribution in testing. Figure 10.9 shows the χ^2-distribution for 5, 10, 15, and 20 degrees of freedom.

The last distribution we want to look at briefly is the *F*-distribution. We need it, for example, when doing analysis of variance and regression. As with the *t*-distribution and the χ^2-distribution, there is not only one *F*-distribution, but many *F*-distributions which depend again on the number of degrees of freedom. The difference is that we have two values for the degrees of freedom. Again, we will deal with this later in the analysis of variance and the regression analysis chapter. Figure 10.10 shows the *F*-distribution for the degrees of freedom 2 and 5 as well as 15 and 15.

For our practical work, it is important to understand that these distributions are theoretical distributions which we need for testing. As we will see later, one or the other

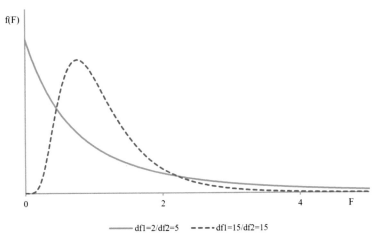

Fig. 10.10 The F-distribution for degrees of freedom df1 equal to 2 and df2 equal to 5, and for degrees of freedom df1 equal to 15 and df2 equal to 15

distribution comes into play depending on the testing procedure. It is also important to know that the total area under the curve is always 1 or 100% and that intervals under the area indicate how likely it is that a particular value of the x-axis will occur, e.g., how likely it is that an F-value greater than 4 will occur. Compare Fig. 10.10, where the area and thus the probability is very small.

10.4   Creating Distributions with Excel

Excel offers the possibility to draw the discussed distributions. The following functions are available for this purpose: NORM.DIST, T.DIST, CHISQ.DIST and F.DIST.

10.4.1 Drawing the Normal Distribution N(100, 20) with Excel

To draw the normal distribution with a mean of 100 and a standard deviation of 20, we first open an empty Excel worksheet and enter the possible x-values in the first column. In the second column, the function NORM.DIST gives the values of the normal distribution back. To do this, we enter the x-value in the NORM.DIST window for which we want to obtain the function value, further we define the mean and the standard deviation and enter

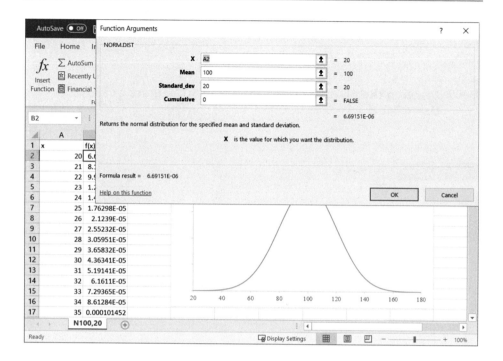

0 for Cumulative. The latter causes us to obtain the probability density function. We then draw the graph using the Insert tab and the Line Chart command.

Drawing the t-distribution, the χ^2-distribution and the F-distribution works similarly. However, instead of the mean and standard deviation, we need to define the degrees of freedom for which we want to obtain the distribution functions. In the following there are three examples:

10.4.2 Drawing the t-Distribution Curve with 15 Degrees of Freedom Using Excel

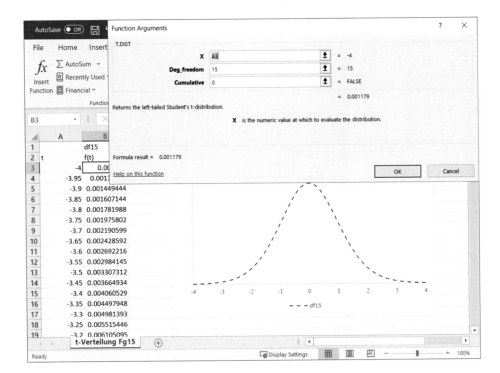

10.4.3 Drawing the χ^2-Distribution Curve with 10 Degrees of Freedom Using Excel

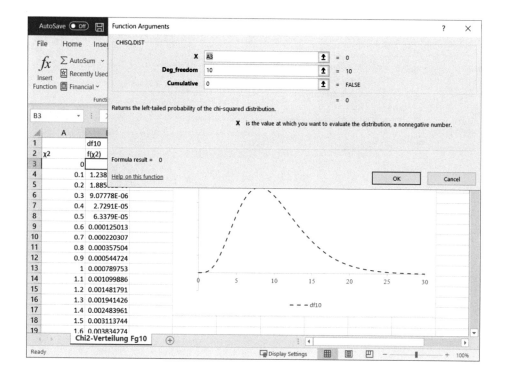

10.4.4 Drawing the F-Distribution Curve with Two Times 15 Degrees of Freedom with Excel

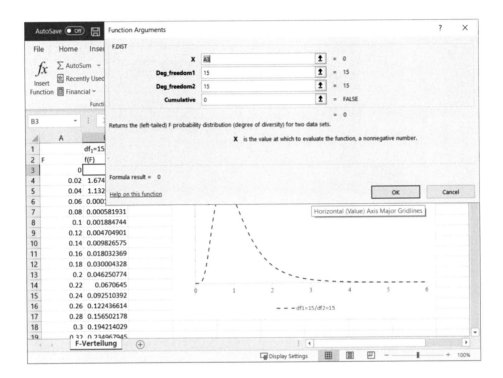

We could also use Excel to calculate interval plots under the distribution curves. However, we will not do so here because the areas are tabulated in Appendices 2–5.

10.5   Checkpoints

- *The normal distribution is bell-shaped, symmetric with a maximum and its tails converge asymptotically to the x-axis.*
- *The area under the normal distribution indicates the probability of finding a value in a particular interval.*
- *Using the z-transformation, we can transform any normal distribution into the standard normal distribution with mean 0 and standard deviation 1.*

- *The standard normal distribution is an important test distribution and its areas and values are tabulated.*
- *Other important test distributions whose values are tabulated are the t-distribution, the χ^2-distribution and the F-distribution.*

10.6 Applications

10.1 Using Excel, draw the normal distribution curves $N(-20, 15)$, $N(20, 15)$ and $N(0, 25)$ and compare them.

10.2 Using the z-transformation, transform the normal distribution $N(20, 15)$ into the standard normal distribution $N(0, 1)$. What are the z values for the following x values: $-25, -20, -10, -7, 5, 20, 35, 47, 50, 60, 65$?

10.3 What is the area under the standard normal distribution curve to the right of $z = 0.5$, 0.75, 1.0, 1.26?

10.4 What is the area under the standard normal distribution curve to the left of $z = -0.5$, $-0.75, -1.0, -1.26$?

10.5 Compare the solutions from Task 3 and Task 4. What do we see?

10.6 Suppose that the age of founders is normally distributed with mean 35 and standard deviation 7.

- What is the probability that we discover a founder who is older than 42 years?
- What is the probability that we discover a founder who is younger than 21 years?
- What is the probability that we discover a founder between the ages of 42 and 49?

10.7 Suppose one of our enterprise founders has developed a novel process to cut panel sheets. The most important thing is the precision of the cutting width. He regularly checks the width and finds that the width of the cuts is 0.1 mm on average, with a standard deviation of 0.005 mm. What is the probability that the cutting width does not meet the tolerance range of 0.0908 mm to 0.1092 mm required by a customer?

10.8 We need to be in the top 5% of students in a selection process to receive a scholarship. We know that the distribution of points on the test is normally distributed, and that there is usually an average of 80 points with a standard deviation of 10.0 points. How many points do we need to score to be in the top 5%?

Hypothesis Test: What Holds?

11

We now know what a hypothesis is and learned about important test distributions. What we do not know yet is how to use test distributions to test hypotheses and make statements about our population. The objective of this chapter is to learn just that. We clarify the concept of statistical significance. Once we get this, we can read scientific studies. We explain the significance level as well as the errors that are possible when testing. We also address the fact that statistically significant is not synonymous with practically relevant. Lastly, we discuss the steps we must go through when conducting any hypothesis test.

11.1 What Does Statistically Significant Mean?

Statistical significance is a central concept in statistics when testing hypotheses. With the help of this concept, hypotheses are rejected or not and statistically significant knowledge is generated. But how does this work? Let's go a step back and start with a hypothesis. As we already know, we formulate a research hypothesis from our research question with the help of existing studies and theories, and from this we formulate our testable null hypothesis H_0 and our alternative hypothesis H_A.

Once again we stress the example of the age of our enterprise founders.

Research question: What is the average age of enterprise founders?

From this we derived the following null and alternative hypothesis via existing studies and theories.

H_0: *Enterprise founders are on average 40 years old.*

H_A: *Enterprise founders are on average younger or older than 40 years old.*

© Springer-Verlag GmbH Germany, part of Springer Nature 2023
F. Kronthaler, *Statistics Applied With Excel*,
https://doi.org/10.1007/978-3-662-64319-8_11

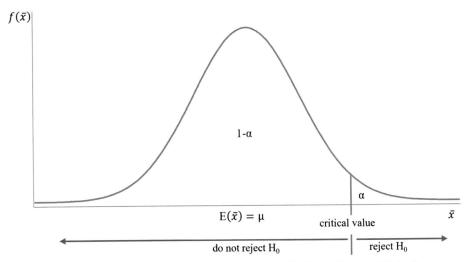

Fig. 11.1 The concept of statistical significance in the one-sided hypothesis test (rejection area on the right)

We can formulate the null and alternative hypothesis a bit shorter.

$$H_0 : \mu = 40$$

$$H_A : \mu \neq 40$$

Let's take a closer look at the short version. H_0 we have described with $\mu = 40$ which is the abbreviated version of the following statement: In the population, enterprise founders are on average 40 years old. μ stands for the mean value of the population. For H_A we wrote $\mu \neq 40$. This means, in the population enterprise founders are not (unequal) 40 years old.

We test our hypotheses with the mean of the sample $\bar{x}$. If our sample result $\bar{x}$ deviates very much from the value we assumed for the null hypothesis and exceeds a critical value, we suppose that our statement about the population is not true. We reject it and proceed with the alternative hypothesis.

But what is a sufficient large deviation or a critical value? The best way to illustrate this is with a figure on the concept of statistical significance using the one-sided hypothesis test. We use this when we have directed hypotheses, but this does not really matter right now (see Fig. 11.1 and Chap. 9).

In the figure we see the distribution of our possible sample means, in short the sampling distribution. We already know that there is not just one sample, but many possible samples and therefore many possible sample means. On the x-axis we have plotted the possible

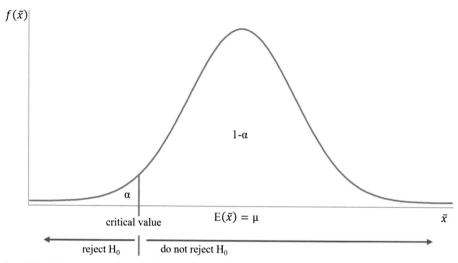

Fig. 11.2 The concept of statistical significance in the one-sided hypothesis test (rejection area on the left)

sample means. On the y-axis we see how often certain sample mean values occur. Further, we already know that the expected value of the sample is equal to the mean of the population: $E(\bar{x}) = \mu$ (compare chap. 8). We see from the figure that values close to the expected value have a high probability to occur, given our null hypothesis is correct. Values far from the expected value have a very small probability. If the probability is too small, that is, if our sample mean exceeds a critical size, then we assume that our null hypothesis cannot be correct and reject it. α is the area of the sampling distribution in the rejection area, i.e. the probability of finding a value that is larger than the critical value.

Of course, it is also possible that the rejection area is on the left-hand side and that we reject the null hypothesis if the sample mean falls below a critical value (compare Fig. 11.2).

For undirected hypotheses, we test two-sided and place our rejection area on both sides, as shown in the following figure (Fig. 11.3).

We divide our area α into two halves, thus have on the left and right $\alpha/2$ and hence on the left a critical value that we can fall below and on the right a critical value that we can exceed. At this point it makes sense to take a closer look at the area of α.

11.2   The Significance Level α

We had already said that α the area under the sampling distribution is the rejection area and represents the probability of finding a value that is greater than the critical value. But how small must the probability be that we reject our null hypothesis? Three areas of

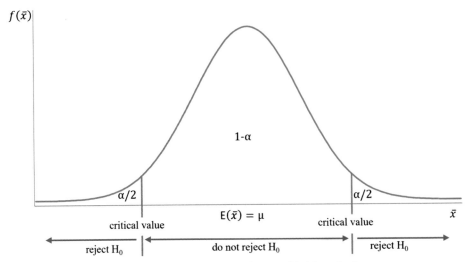

Fig. 11.3 The concept of statistical significance in the two-sided hypothesis test

different sizes have become widely accepted in science: $\alpha = 10\%$ *resp.* $\alpha = 0.1$, $\alpha = 5\%$ *resp.* $\alpha = 0.05$ and $\alpha = 1\%$ *resp.* $\alpha = 0.01$. That is, we reject the null hypothesis if the probability of finding a particular value is less than 10%, 5%, or 1%. Instead of 10%, 5% or 1% we can also write 0.1, 0.05 or 0.01, we use then the decimal notation instead of percentages. We refer to these areas as the significance level α. If our sample yields a value whose probability is less than 10%, 5% or 1%, then we speak of rejecting the null hypothesis at the significance level of 10%, 5% or 1%. The result is statically significant. In other words, the probability of finding a sample value at this level is smaller than 10%, 5%, or 1% given the null hypothesis is correct. α is thus the probability at which we reject our null hypothesis.

But there is more to say about α. α is also the error probability with which we falsely reject the null hypothesis. Look again at Fig. 11.3. On the x-axis all sample means are plotted that are possible under given the null hypothesis is true. Some are more likely, some are less likely. Values to the left and right are unlikely, but not impossible. But if we find a value far to the left or far to the right, we reject the null hypothesis, knowing well that the value is possible given the null hypothesis is true. That is, there is a possibility that we make an error in our test decision. This error probability is our chosen significance level α.

This possible error should be considered when we choose our α. If the consequences of a false rejection are serious, if life or death, so to speak, depends on it, then we choose a small α, α equal to 1% or perhaps even smaller, like α equal to 0.1%. If the consequences are less severe, we allow a larger α, perhaps α equal to 5% or 10%. Let's illustrate this again with an example. A bridge builder says that his bridge, over a gorge, holds with a certainty of 90% when we walk over it, or that the probability of collapsing is only

Table 11.1 α-error and β-error in the hypothesis test

	H_0 is true	H_0 is not true
H_0 is rejected	α-error	correct decision
H_0 is not rejected	correct decision	β-error

10%. Are we going to walk over this bridge? Probably not, the risk (probability) of the bridge collapsing would have to be much lower. Let's take a second example. The weather forecast says that there is a 5% chance of rain tomorrow. Will we leave the umbrella at home? Very likely, if it nevertheless rains, we'll just get wet, but nothing else will happen. The important thing here is to think about the implications of a false rejection of the null hypothesis in advance of the hypothesis test, and to choose our α according to these considerations.

But why do we choose our α not just as small as possible? There are two reasons for this. First, the rejection of the null hypothesis becomes less and less likely the smaller our α is. Second, there is not only the problem of a false rejection of our null hypothesis, but we might also fail to reject the null hypothesis even though it is false. This is the second possible error in hypothesis testing, the β-error. In Table 11.1) both errors are shown.

This table is so important that we want to discuss it in more detail. The first row tells us whether our null hypothesis actually reflects the state of the population, that is, whether H_0 is true or false. We do not know this. Our null hypothesis is merely a guess about the behavior in the population. Accordingly, our null hypothesis can be both true and false. The second and third rows contain the result of the hypothesis test. Let's look at the second row first. The result of the hypothesis test is that we reject our null hypothesis. This is true if our null hypothesis is indeed false. However, if our null hypothesis correctly reflects the behavior in the population, then we have made a false test decision, the α-error. We have wrongly rejected the null hypothesis. Now let's look at the third row. The result is that we do not reject the null hypothesis. This is true if our null hypothesis is correct. But if our null hypothesis does not correctly reflect the behavior in the population, we have made an error called the β-error.

It is now important that the α-error and the β-error are not independent of each other. If we reduce the α-error, we increase the β-error and vice versa. In Fig. 11.4 this relationship is graphically displayed. If we decrease the area of the α-error in this figure, we increase the area of the β-error and vice versa.

It is also important for practical work that the β error is relatively difficult to determine. To determine the β-error, we need to know how to formulate the null hypothesis correctly so that it represents the population. But we do not have this knowledge, otherwise we would not have formulated the null hypothesis as it is. For this reason, in practice we usually work with the α-error.

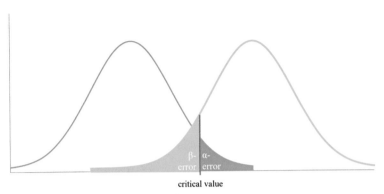

Fig. 11.4 Relationship between α-error and β-error

However, we can still draw an important conclusion from what has been said. If we perform the hypothesis test and reject our null hypothesis or not there is a probability that we have made a mistake, even if we have worked by all rules. It follows that a test decision is always only an indication of the behavior of the population, but never an absolute proof.

11.3   Statistically Significant, But also Practically Relevant?

We have seen how we come to decide whether a result is statistically significant and what this all about. We can say with a relatively high level of confidence that a result is not random. But is such a statistically significant result always practically relevant or practically significant? This must be clearly answered with no. Statistically significant does not automatically mean practically relevant. The background is that the sample size has a large influence on whether a result becomes statistically significant. If the sample is large enough, every result becomes statistically significant, even very small differences which may be practically completely insignificant.

Let's illustrate this with a small fictional example. Let's imagine we are interested in beer consumption at the Oktoberfest in Munich (or at any other folk festival). Specifically, we analyze whether there is a difference in beer consumption between male and female folk festival visitors. Let's assume we collect data on this and find that during a two-hour visit men drink an average of 1.98 litre of beer (almost two Maßbeer) and women drink 1.94 litre (almost two Maßbeer as well). Now, if the sample is sufficiently large, then this result becomes statistically significant. But is this result also practically relevant? Probably not, a difference of 0.04 liters does not matter with these quantities, it is in any case too much beer. Just to mention, there are really people that drink that much at the Oktoberfest.

We see that a statistically significant result should always be analyzed for its practical significance.

We can evaluate this using absolute numbers and measures of effect size. Measures of effect size are standardized numbers that provide information about how strong the effect is. A major advantage of calculating effect sizes is that results from different studies can be compared.

Commonly used measures of effect size are the Bravais-Pearson correlation coefficient r and *Cohen's d*. How to calculate the measures are discussed in the next chapters, when we perform the test procedures. However, a note on the interpretation of the effect size should be given here. It is generally accepted that the magnitude of the effect size r can be interpreted as follows:

an effect size of $r = 0.1$ indicates a small effect,
an effect size of $r = 0.3$ is a medium effect,
an effect size of $r = 0.5$ and larger measures a large effect.

In summary, in addition to statistical significance, practical relevance should be analyzed using effect size and absolute numbers.

11.4   Steps When Performing a Hypothesis Test

Before we now turn to the hypothesis test, let us briefly go through the five steps that should to be carried out when doing a hypothesis test:

1. Formulating the null hypothesis, the alternative hypothesis, and define the significance level
2. Determining the test distribution and test statistic
3. Determine the critical value and the rejection area
4. Calculating the test statistic
5. Decision and interpretation

1. *Formulating the null hypothesis, the alternative hypothesis, and define the significance level* The first step in hypothesis testing is always to formulate the research hypothesis from the research question and from that the null hypothesis and the alternative hypothesis. It cannot be stressed enough, that the hypotheses cannot be formulated from the gut, but must reflect the existing knowledge, theories and studies. In addition, we have to determine at what significance level we are willing to reject the null hypothesis. The choice of the significance level depends as already discussed on the effects of an incorrectly rejection of the null hypothesis.
2. *Determining the test distribution and test statistic* We learned about different test distributions in Chap. 10. Depending on the null hypothesis, we have different test situations associated with different test distributions and test statistics. Once we have learned about the different test situations, we can automatically associate the null

hypothesis with a specific test distribution and test statistic. This will be discussed in the following chapters.

3. *Determine the critical value and the rejection area* In Chap. 10 we have seen that the area under the test distributions indicates how likely a particular value and larger is to occur. Having set the significance level and knowing the test distribution (standard normal distribution, t-distribution, χ^2-distribution or F-distribution), we are able to determine the critical value and hence the rejection area.

4. *Calculating the test statistic* The next step is then to calculate the test statistic. For this, we compare the sample value found with the value expected under the null hypothesis. As already mentioned the test statistic depends on the test situation.

5. *Decision and interpretation* The last step is to make the test decision and to interpret the results. Here we compare the test statistic with the critical value. If the value of our test statistic falls within the rejection area, we reject the null hypothesis at our significance level and we continue to work with the alternative hypothesis. It is important to note here that our test decision never can be a proof of the null hypothesis or the alternative hypothesis. However, we have a good indication of how the population is likely to behave.

11.5   How do I Choose My Test?

Above we mentioned that there are different test situations. They depend on what we want to know and on the scale level of the data. Figure 11.5 shows important test situations and illustrates which test procedure is used in which situation.

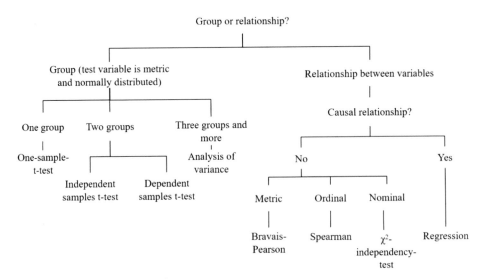

Fig. 11.5 Selecting a test procedure

First it is important whether we are examining groups or relationships between variables. For groups, we go along the left axis and arrive at the one-sample test, the independent-samples test, and the dependent-samples test. If we analyze relationships between variables, we go along the right axis and arrive at the various correlation coefficients or regression, depending on the question of causality. We are now ready to begin. In the following chapter we will start with our first test situation, the one-sample t-test.

11.6 Checkpoints

- The significance level α is the probability of finding a value greater than a critical value, and thus the probability or certainty at which the null hypothesis is rejected.
- α is usually chosen at the level of 10%, 5% or 1%. Alternatively $\alpha = 0.1$, $\alpha = 0.05$ and $\alpha = 0.01$ is written.
- The choice of α depends on what happens if we falsely reject the null hypothesis. If the consequences are severe, we choose a smaller α. If the effects are less severe, we choose a larger α.
- There are two possible errors when doing a hypothesis test. The α-error is the probability of rejecting the null hypothesis even though it is true. The β-error is the probability of not rejecting the null hypothesis even though it is not true.
- α-error and β-error are related to each other. Decreasing one error increases the other and vice versa.
- A statistically significant result should always be analyzed for its practical significance.
- There are five steps to be performed when doing a hypothesis test: (1) formulating the null hypothesis, alternative hypothesis, and defining the significance level, (2) determining the test distribution and test statistic, (3) determining the critical value and the rejection area, (4)calculating the test statistic, and (5) deciding and interpreting.
- The choice of the test procedure depends on the test situation and the scale of the data.

11.7 Applications

11.1 What are the two meanings of the significance level α?

11.2 Which of the following statements is correct:

- Exceeding the critical value leads to the rejection of the null hypothesis.
- It is possible to set the α-error to zero.
- The smaller the α-error the better the result.
- The selection of the α-error depends on the effects of falsely rejecting the null hypothesis.

11.3 When to test two-sided, left-sided or right-sided?

11.4 Formulate both the null hypothesis and the alternative hypothesis for the two-sided test, the left-sided test, and the right-sided test for the following research question: What is the age of enterprise founders?

11.5 What is the relationship between the α-error and the β-error?

11.6 Go to the library and find an article of interest that uses data to generate knowledge about the population. Explain how the author used the concept of statistical significance to generate knowledge about the population.

Time to Apply the Hypothesis Test

Now we start with hypothesis testing and to generate information about the population. If our information about the 100 enterprises were data from a real sample, we would now generate information about all start-up enterprises with the help of the sample data.

We start with the test for a group mean and the test for a difference between group means for metric data. These test procedures assume that the test variable is metric and normally distributed in the population. The requirement of normally distributed data is particularly important for small samples. For large samples, the tests are relatively robust to a violation of this assumption. We then discuss analysis of variance, and test procedures for correlation between metric, ordinal, and nominal variables. Finally, we learn about further test procedures for nominal data.

The Test for a Group Mean or One-Sample t-Test 12

12.1 Introduction to the Test

Remember, in Chap. 3 we learned about and calculated the arithmetic mean. For example, we found out that the enterprises have grown on average by 7.1% in the last 5 years, that they spend on marketing on average 19.81% and on product improvement on average 4.65% of the turnover. Furthermore, we know that the enterprise founders are on average 34.25 years old and that they have a work experience of 7.42 years. We have learned that the statement only applies to the sample, not to the population. However, we are not satisfied with this. We can use the hypothesis test to generate a statement about the population. Specifically, we use the test for a group mean to determine whether the assumption about the mean in the population is in line with the sample findings.

12.2 The Research Question and Hypothesis: Are Company Founders on Average 40 Years Old?

In order to test the assumption about the mean in the population, we must first formulate the assumption. As shown above, we specify the research question and formulate our hypothesis using existing studies and theories. Let's take the example of the age of enterprise founders. The research question could be then: What is the average age of enterprise founders? Using theories and studies, we generate our null hypothesis and our alternative hypothesis:

H_0:   *Enterprise founders are on average 40 years old.*
H_A: *Enterprise founders are on average younger or older than 40 years old.*

© Springer-Verlag GmbH Germany, part of Springer Nature 2023
F. Kronthaler, *Statistics Applied With Excel*,
https://doi.org/10.1007/978-3-662-64319-8_12

respectively

H_0: $\mu = 40$
H_A: $\mu \neq 40$

In addition, the significance level should be defined at which we reject the null hypothesis. An incorrect rejection of the null hypothesis does not seem to be a matter of life or death. At the same time, we maybe want to be reasonably sure. Hence, we set the significance level at 5%:

$$\alpha = 5\% \quad \text{or} \quad \alpha = 0.05$$

Again, as a reminder. A significance level of 5% means that we reject the null hypothesis if we obtain a sample finding that has a probability to be found less than 5% when the null hypothesis is correct.

12.3   The Test Distribution and Test Statistic

Now we can move on to the second step in the hypothesis test. As we have already heard, each test is associated with a particular test distribution and test statistic. In the test for a group mean with a large sample $n > 30$ the test distribution is the standard normal distribution (when the standard deviation of the population is known). The test statistic compares the sample mean to the mean assumed in the null hypothesis:

$$z = \frac{\bar{x} - \mu}{\sigma_{\bar{x}}}$$

with

z is the z-value of the standard normal distribution,
$\bar{x}$ is the sample mean,
μ is the population mean assumed in the null hypothesis,
$\sigma_{\bar{x}}$ is the standard deviation of the sampling distribution with $\sigma_{\bar{x}} = \frac{\sigma}{\sqrt{n}}$,
σ is the standard deviation in the population, and
n is the number of observations.

12.4   The Critical Value

We determine the critical value, and thus the rejection area, using the standard normal distribution table. To do this, we need to know whether we test one-sided or two-sided and what our significance level is. When the test is one-sided, we place the significance level either on the left or the right side. When the test is two-sided, we split the significance level and place the two half's on the left and on the right side. In our example, the test is two-sided, because we have an non-directional null hypothesis, at the 5% significance level. That is, we have an area of 2.5% on both the left and the right side with the critical z-values $-$ 1.96 and 1.96 (compare Appendix B).

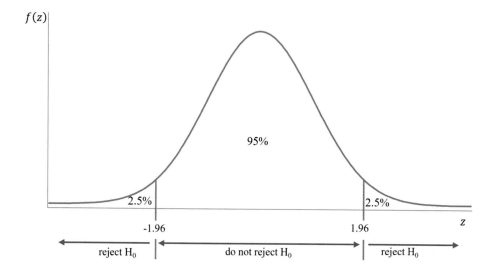

12.5   The z-Value

Now we just need to calculate the z-value. We do this using the test statistic. We insert the sample mean, the mean from the null hypothesis, and the standard deviation of the sampling distribution. As we calculated earlier, the mean of the sample is $\bar{x} = 34.25$ years. For the mean of the null hypothesis, we insert the assumed value of $\mu = 40$ years. To calculate the standard deviation of the sampling distribution, we need the standard deviation in the population. Here let's assume that it is 8 years, accordingly we obtain the following value: $\sigma_{\bar{x}} = \frac{\sigma}{\sqrt{n}} = \frac{0.8}{\sqrt{100}} = 0.80$ years. Our z-value is thus:

$$z = \frac{\bar{x} - \mu}{\sigma_{\bar{x}}} = \frac{34.25 - 40}{0.80} = -7.19$$

12.6   The Decision

The only thing missing now is the decision. For this we compare our z-value with the critical values. We see that our z-value is to the left of the critical value -1.96, i.e. we reject the null hypothesis and come to the interpretation. We can now say that the enterprise founders in the population are not 40 years old in average. Our sample mean also suggests that enterprise founders are younger than 40 years old. We could now compute the confidence interval, as we did in Chap. 8. For example, if we determine the 95% confidence interval, we come to the conclusion that the true value lies between 32.68 and 35.82 years with a probability of 95%.

12.7   The Test When the Standard Deviation in the Population Is Unknown or the Sample Is Small n ≤ 30

We tend to have the problem that we do not know the standard deviation in the population. In this case, we substitute in the formula of the standard deviation of the sampling distribution

$$\sigma_{\bar{x}} = \frac{\sigma}{\sqrt{n}}$$

the standard deviation of the population σ by the standard deviation of the sample s. Thus, the standard deviation of the sampling distribution is recalculated as follows:

$$\hat{\sigma}_{\bar{x}} = \frac{s}{\sqrt{n}}$$

with

$\hat{\sigma}_{\bar{x}}$ is the estimated standard deviation of the sampling distribution,
s is the standard deviation of the sample, and
n is the number of observations.

The test distribution is then no longer the standard normal distribution, but the t-distribution with $df = n - 1$ degrees of freedom. The same is the case when we have a small sample. We speak of a small sample in this test when the number of observations n is less than or equal to 30 ($n \leq 30$). In both cases, the critical values are not taken from the standard normal distribution table, but from the t-distribution table (Appendix C). Consequently, the test value is t-distributed and the test statistic is as follows:

$$t = \frac{\bar{x} - \mu}{\hat{\sigma}_{\bar{x}}}$$

Apart from that, the procedure is identical. Let's demonstrate this using a small sample as an example. We assume that we have only 20 observations (n = 20) and take the sample mean and standard deviation from our data set. The values are $\bar{x} = 31.55$ and $s = 7.96$. We also want to test one-sided. To test one-sided, we need to reformulate the hypotheses. The null hypothesis and the alternative hypothesis should be as follows.

H_0: *Enterprise founders are on average older than or equal to 40 years.*
H_A: *Enterprise founders are on average younger than 40 years.*

respectively

H_0: $\mu \geq 40$
H_A: $\mu < 40$

We leave the significance level at $\alpha = 5\%$ resp. $\alpha = 0.05$.
The test distribution is t-distributed with $df = n - 1 = 20 - 1 = 19$ degrees of freedom and the test statistic is:

$$t = \frac{\bar{x} - \mu}{\hat{\sigma}_{\bar{x}}}$$

The rejection area at the 5% significance level is on the left. We reject the null hypothesis if we fall below a certain value. We take the critical value from the t-distribution table, it is -1.729 (see Appendix C).

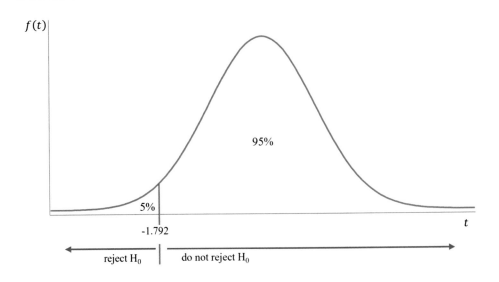

The test statistic is as follows:

$$t = \frac{\bar{x} - \mu}{\hat{\sigma}_{\bar{x}}} = \frac{31.55 - 40}{1.78} = -4.75$$

If we now compare our calculated t-value with the critical value, we see that we fall short of it and that we are in the rejection area. That is, we reject the null hypothesis and continue to work with the alternative hypothesis. So, on average, our enterprise founders are younger than 40 years.

Freak Knowledge
We just saw that for a small sample, the test value is t-distributed and the test distribution is the t-distribution. In Chap. 10 we learned that for a large number of observations the t-distribution becomes the standard normal distribution. In statistical programs, therefore, only the t-distribution is used as the test distribution, and the test for group means is often called the one-sample t-test.

12.8   The Effect Size

Now, before we turn to the final interpretation, let's calculate the effect size. The measure usually used with the one-sample t-test is *Cohen's d*. It is calculated as follows:

$$Cohen's\ d = \frac{\bar{x} - \mu}{s}$$

with

$\bar{x}$ is the arithmetic mean of the sample,
μ is the population mean assumed under the null hypothesis, and
s is the standard deviation of the sample.

If we look at the formula, the only difference from the test statistic is that we are not dividing by the standard deviation of the sampling distribution (standard error), but by the standard deviation of the sample. What does this mean? When we divide by the standard deviation, we calculate the deviation in terms of the standard deviation. Thus, a Cohen's d of 1 means that the deviation is one standard deviation. In the literature, it is suggested to interpret Cohen's d as follows:

- small effect, Cohen's d is in the range between 0.2 to <0.5,
- medium effect, Cohen's d ranges from 0.5 to <0.8,
- large effect, Cohen's d is ≥0.8.

Note Cohen's d can also take negative values, this is the case when the arithmetic mean is smaller than the assumed mean in the population.

Calculating the effect size for our example hence yields a large effect:

$$Cohen's\ d = \frac{\bar{x} - \mu}{s} = \frac{31.55 - 40}{7.96} = -1.06$$

All in all we can make the final interpretation approximately as follows. The test shows that the sample mean is statistically significant different from the assumed mean in the null hypothesis. We reject the null hypothesis at the 5% significance level and proceed with the alternative hypothesis. The effect is large and is more than eight years in absolute terms. Moreover, we could additionally report the confidence interval, which would be even better.

Freak Knowledge
The one-sample t-test assumes that the data are metric and normally distributed. But what do we do if the data is metric but not normally distributed or the data is ordinal. In such a case, we have the Wilcoxon test. We use the Wilcoxon test mainly when we have a small sample and non-normally distributed metric data. The problem of testing an average value for ordinal data (we take the median) occurs rather rarely.

12.9　Calculating the One Sample t-Test with Excel

Excel does not provide a function for the one-sample t-test. We have to calculate the test statistic by hand as shown above. However, we can use the function NORM.S.DIST to calculate the exact probability with which our test value occurs. We will briefly demonstrate this using our example and the calculated test value of -7.19. If we call the function, enter the value of the test statistic as the z-value and 1 under Cumulative, we get back from Excel the probability of the area left of the test statistic. As we can see, the probability is very small. That is, the probability of obtaining such a low value when the null hypothesis is correct is much smaller than 2.5%. As a comparison, for clarity, the z-value of -1.96 and the corresponding area are also shown. As we can see from our table of the standard normal distribution (Appendix B), the area to the left of this z-value is 2.5%.

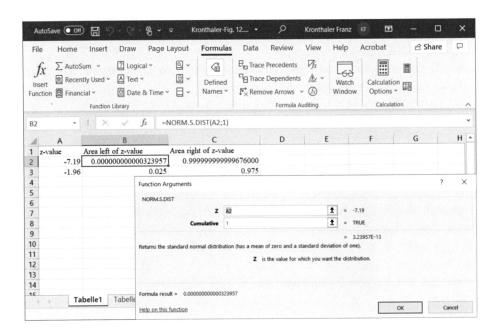

Accordingly, we can use the function T.DIST to calculate the exact probability of our t-value to occur when the null hypothesis is correct. If we call the function, enter our t-value, define the number of degrees of freedom and enter 1 under Cumulative, we get from Excel the probability of the area to the left of the t-value.

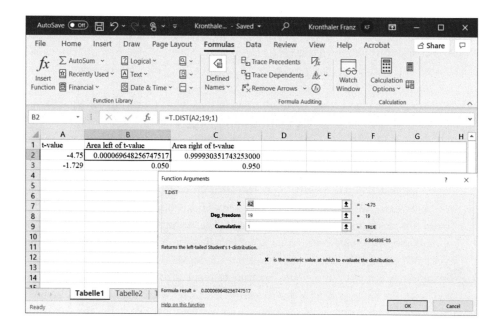

12.10 Checkpoints

- *In the test for a group mean, we address the question of whether the assumed mean of the population is in line with our sample finding.*
- *If the sample is large and we know about the standard deviation in the population, the test distribution is the standard normal distribution and our test value is standard normal distributed.*
- *If the sample is small or if we do not know the standard deviation in the population, the test distribution is the t-distribution and our test value is t-distributed.*
- *Cohen's d is an often used measure to calculate the effect size.*
- *The test for a group mean is often referred to as the one-sample t-test.*

12.11 Applications

12.1 We are interested in the professional experience that enterprise founders have on average. Based on prior considerations and studies we read, we hypothesize that they have typically accumulated ten years of professional experience before starting an enterprise. Accordingly, we formulate the following null hypothesis: The average professional experience is 10 years. Test the null hypothesis at the 10% significance level considering all relevant steps.

12.2 One of our companies has grown by an average of 5% in recent years. We are now interested in whether the company has grown below average or above average compared to the population. To test this, we formulate the following null hypothesis: The average growth of young enterprises in the population is 5%. Test the null hypothesis at the 5% significance level considering all relevant steps.

12.3 Assume that our sample contains only the first 25 firms in our data set. Perform task 2 again for this problem.

12.4 We are interested in whether the proportion that firms spend on marketing has fallen in recent years. We know that ten years ago the expenditure was in average 15%. To test this, we formulate the null hypothesis: The average turnover proportion spent on marketing of young enterprises is greater than or equal to 15%. Test the null hypothesis at the 10% significance level considering all relevant steps.

12.5 We know that ten years ago, young enterprises spent on average 5% of their turnover on innovation. We hypothesize that this effort has increased as the literature discusses the increasing importance of innovation. We want to test this using the following null hypothesis: The average turnover proportion spent on innovation of young enterprises is less than or equal to 5%. Test the null hypothesis at the 5% significance level considering all relevant steps.

The Test for a Difference Between Group Means or Independent Samples t-Test   13

13.1 Introduction to the Test for Difference Between Group Means with Independent Samples

In the last chapter we explored the question of how old our enterprise founders are on average and we conducted a test for a group mean. We can extend this question. For example, we can be interested in whether women are older than men when they start a business. In this case, we compare the average start-up age of women with the average start-up age of men. We compare two means and test whether they differ or whether they are the same. Whenever we compare two means from two different groups, we perform the test for a difference between group means with independent samples. Other examples from our data set would be the questions of whether industrial firms or service firms achieve higher average growth or whether women and men differ in terms of job experience.

13.2 The Research Question and Hypothesis: Are Women and Men of the Same Age When Starting an Enterprise?

First, as always, the research question has to be specified. The research question could be as follows. Are women and men of the same age when starting an enterprise? To formulate the null hypothesis and the alternative hypothesis, we first read the literature. Using this, we may conclude that women and men differ in the average age when starting an enterprise. Thus, our null hypothesis is as follows:

H_0:   *Male and female founders are of the same age when they start an enterprise.*
H_A: *Male and female founders are not of the same age when they start an enterprise.*

© Springer-Verlag GmbH Germany, part of Springer Nature 2023
F. Kronthaler, *Statistics Applied With Excel*,
https://doi.org/10.1007/978-3-662-64319-8_13

We can write this more briefly:

$$H_0 : \quad \mu_1 - \mu_2 = 0$$
$$H_A : \quad \mu_1 - \mu_2 \neq 0$$

Translated, means this for the null hypothesis, the difference between the mean of group one μ_1 (e.g. women) and the mean of group two μ_2 (e.g. men) is in the population zero. That is, the means are equal in the population. The alternative hypothesis states that the difference between the means of the respective groups is not equal to zero. Additionally, we specify the significance level as 10%.

$$\alpha = 10\% \quad \text{or} \quad \alpha = 0.1$$

Now we can run our test.

13.3   The Test Distribution and the Test Statistic

The test distribution is the t-distribution with $df = n_1 + n_2 - 2$ degrees of freedom, where n_1 is the number of observations of group one and n_2 the number of observations of group two. The test statistic is:

$$t = \frac{\bar{x}_1 - \bar{x}_2}{\hat{\sigma}_{\bar{x}_1, \bar{x}_2}}$$

where

t	is the t-value of the t-distribution,
$\bar{x}_1$ and $\bar{x}_2$	are the sample means of the respective groups,
$\hat{\sigma}_{\bar{x}_1, \bar{x}_2}$	is the estimated standard deviation of the sampling distribution.

We see, the test statistic compares the two group means. If the two means are equal, the upper part of the equation, the numerator, is zero and the t-value is also zero. The larger the difference in the means, the larger the t-value becomes. We now need to take a closer look at the standard deviation of the sampling distribution. It is estimated as follows:

$$\hat{\sigma}_{\bar{x}_1, \bar{x}_2} = \sqrt{\left[\frac{(n_1 - 1)s_1^2 + (n_2 - 1)s_2^2}{n_1 + n_2 - 2}\right]\left[\frac{n_1 + n_2}{n_1 \times n_2}\right]}$$

where

n_1 and n_2 is the number of observations of the respective groups,
s_1 is the standard deviation of the sample of the first group,
s_2 is the standard deviation of the sample of the second group.

Admittedly, the formula looks a bit complicated, but as with the test for one group mean, it only includes the standard deviations and the number of sample observations. We have knowledge know about the test statistic and the test distribution, so we can turn our attention to the critical value and the rejection area.

13.4   The Critical t-Value

We determine the critical value using the t-distribution table and the degrees of freedom. As with the mean test, it is important to decide whether we want to test one-sided or two-sided and at what significance level. In our example, we test two-sided and at a significance level of 10%. That is, we have an area of 5% on both sides. We take the critical values from the t-distribution table using our degrees of freedom. If we let our male founders be group 1 and our female founders be group 2, then $n_1 = 65$ and $n_2 = 35$, we have 65 men and 35 women in the sample. Thus we have $df = n_1 + n_2 = 65 + 32 - 2 = 98$ degrees of freedom. Knowing this, we obtain the critical values at -1.66 and 1.66 (compare Appendix C).

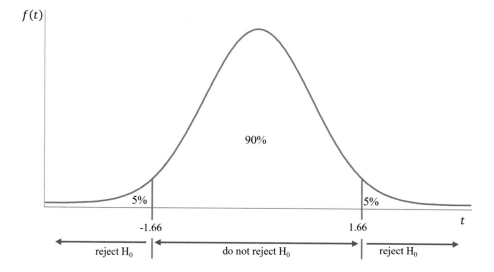

13.5   The t-Value and the Decision

Now we only need to calculate the t-value, we fill in our test statistic with the required values. Calculating the means and standard deviations for our founders should not be a problem. We insert the values and get a t-value of 0.93.

$$t = \frac{\bar{x}_1 - \bar{x}_2}{\hat{\sigma}_{\bar{x}_1, \bar{x}_2}} = \frac{\bar{x}_1 - \bar{x}_2}{\sqrt{\left[\dfrac{(n_1 - 1)s_1^2 + (n_2 - 1)s_2^2}{n_1 + n_2 - 2}\right]\left[\dfrac{n_1 + n_2}{n_1 \times n_2}\right]}}$$

$$= \frac{34.77 - 33.29}{\sqrt{\left[\dfrac{(65 - 1) \times 7.55^2 + (35 - 1) \times 7.76^2}{65 + 35 - 2}\right]\left[\dfrac{65 + 35}{65 \times 35}\right]}} = \frac{1.48}{1.598} = 0.93$$

Using the t-value, we can make the test decision. A comparison with the critical values shows that we do not reject the null hypothesis. So we go on with the null hypothesis.

13.6   The Effect Size

Now before we come to the final interpretation, let's see how we can calculate and interpret the effect size. In this case, it would not be necessary to calculate the effect size, since we already assume that there is no difference in the population. However, if we have a significant result, we should address it.

A commonly used measure of the effect size in the independent samples t-test is the Bravais-Pearson correlation coefficient r. Compared to Cohen's d, it has the advantage of having a fixed range of values from -1 to $+1$. To calculate the effect size in this way, we can convert the t value found into the correlation coefficient. The formula for this is:

$$r = \sqrt{\frac{t^2}{t^2 + df}}$$

where

t is the t-value of the test statistic and
df is the degrees of freedom.

The values can be interpreted as follows:

- a r smaller < 0.3 indicates a small effect,
- a r in the range between 0.3 to < 0.5 is a medium effect,
- a r in the range of 0.5 to 1.0 indicates a large effect.

Since r cannot be greater than 1.0, 1.0 is of course also the theoretically possible maximum effect.

If we put the values from our example into the formula, we get a value that indicates no effect:

$$r = \sqrt{\frac{t^2}{t^2 + df}} = \sqrt{\frac{0.93^2}{0.93^2 + 98}} = 0.09$$

Thus, we can make the interpretation. The result of the independent samples t-test is not significant, i.e. we do not expect a difference in the average age between women and men in the population based on the test result. The effect size also indicates no effect, and the difference in the sample is relatively small, about 1.5 years (compared to the average age of founding).

> **Freak Knowledge**
> In the test procedure just described, the prerequisite, as was mentioned, is a metric and normally distributed test variable. If the test variable is metric but not normally distributed, we can also use the described test procedure for a large sample. The test is then relatively robust against a violation of the assumption of normally distributed data. For metric but not normally distributed data and a small sample, as well as for ordinal data, we can perform the Mann-Whitney test.

13.7   Equal or Unequal Variances

In the example just shown, we simply assumed that the variation in the variable age is the same for both groups, male and female founders. However, we can of course check this, for example with the help of the boxplot (cf. Chap. 4).

According to the boxplots, the variation seems to be slightly higher for women, both, for the interquartile range and the range. The same is seen when we compare the variance, this is 57.06 for male founders and 60.27 for female founders. With some experience it is easy to say that the difference in this case is relatively small and the variances are roughly equal in the population.

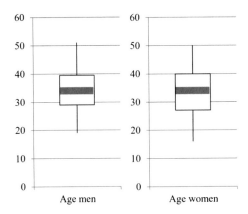

Age men   Age women

Freak Knowledge
Of course, we can also test whether the variances are equal using Levene's test for equality of variances. The test works with the null hypothesis, the variances from both groups are equal, and the alternative hypothesis, the variances from both groups are not equal.

However, the variance among the groups studied need not always to be equal. What is important then is that the test procedures differ slightly with equal or unequal variances. We will see, Excel provides the possibility to compute the test with equal or unequal variances. That is, before we run our test with Excel, we should look at the group variations. If we are unsure about having equal or unequal variances, we simply can calculate the test with Excel for both situations and hope that the results do not differ. If they do, we need to look into the question more deeply.

13.8   Calculating the Independent Samples t-Test with Excel

As already said, we can calculate the test for a difference between group means using Excel. The commands are available within the Data tab and in the Data Analysis tool of Excel: *t*-Test: Two-Sample Assuming Equal Variances and *t*-Test: Two-Sample Assuming Unequal Variances.

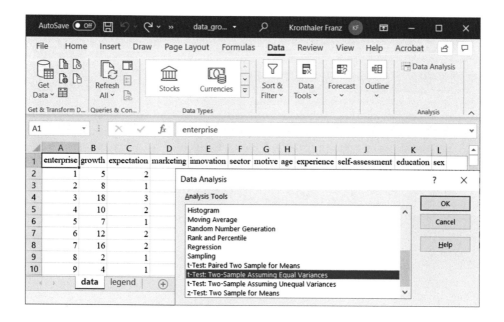

Let's carry out the test for our example above. The best way to do this is first to sort our data by sex. If you don't know how to do this, then consult Excel's help function. After sorting, we may have the zeros at the top (men), the ones are further down (women).

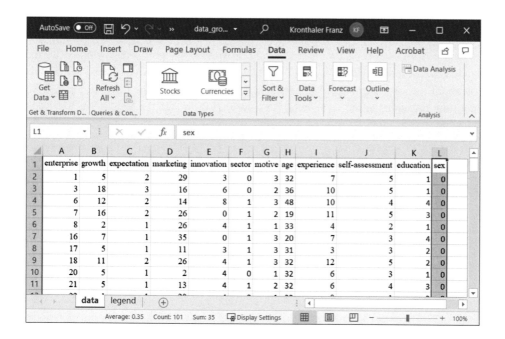

Next, we click on the Data tab, then on Data Analysis, finally on the command t-Test: Two-Sample Assuming Equal Variances, the input screen opens.

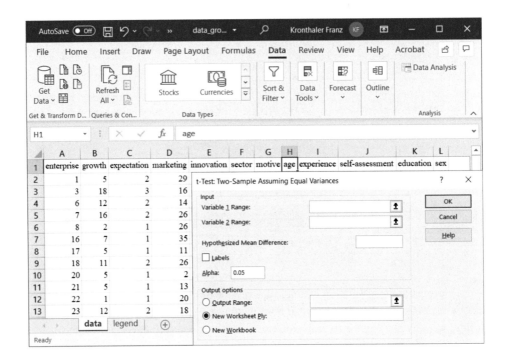

Now we have to fill in the necessary data. We enter for variable *1* the values of one group, e.g. men, and for variable *2* the values of the other group, e.g. women. In the Hypothesized Mean Difference field, we enter zero. In our null hypothesis, we assume that the difference is zero. For Alpha, we enter our significance level. For example, if we are testing two-sided at the 10% level, we enter 0.1. In the Output Range, we specify at which place we want to receive the test results. For example, we specify a new worksheet and enter a name for it.

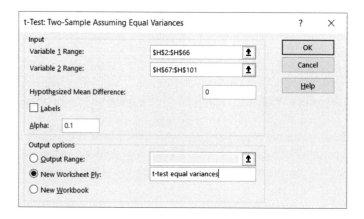

If we click OK, we get the results of our test in the defined place.

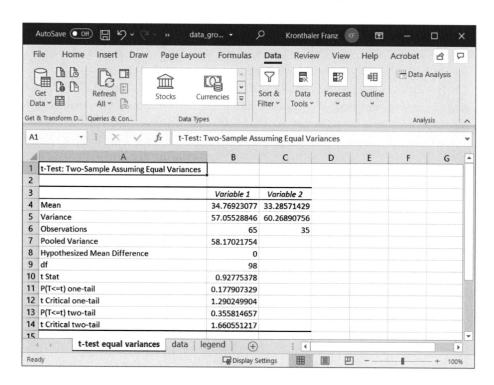

If we compare the output with our example from above, most of the numbers look familiar. In row 4 we see the sample means for group 1 and group 2, which in our case is the average age of the male and female founders. Below in line 5 is the variance for the two groups and once again below in line 6 is the number of observations. We have 65 male founders in the sample and 35 female founders. We can safely ignore line 7 "Pooled

Variance". In line 8 we find the difference assumed under the null hypothesis. Line 9 gives our degrees of freedom back and in line 10 the t-value of test statistic is displayed with 0.928. Now only the four bottom lines remain. In lines 12 and 14 we find the critical t-values for the one-sided and two-sided test. In line 11 and 13, we see the probability for the one-sided and two-sided test of obtaining our calculated t-value and greater, assuming the null hypothesis is correct. P stands for probability. We focus on line 13 because we tested two-sided. We see that the probability of obtaining the calculated t-value of 0.928 when the null hypothesis is correct is to 0.356 or 35.6%. The probability is thus much higher than our specified significance level of 10%. That is, we do not reject the null hypothesis. We also come to this conclusion when we compare the calculated t-value with the critical t-values.

We are now able to interpret the Excel output and to perform the test for a difference between group means for metric and normally distributed data using Excel.

13.9   Checkpoints

- *In the test for a difference between group means, we address the question whether two groups in their population differ with respect to the average behavior.*
- *The test assumes that the test variable is metric and normally distributed.*
- *If the sample is small and the test variable is metric but not normally distributed, or if the test variable is ordinal, we can use the Mann−Whitney test.*
- *The test for a difference between group means is often called independent samples t-test.*
- *Depending on whether the variances are equal or unequal, we perform the test for equal variances or unequal variances.*

13.10   Applications

13.1 We are interested in whether men and women are of the same age when they start a business. Theoretical considerations and previous studies lead us to assume that men are older than women on average. Perform the test by hand at the 10% significance level, taking into account all relevant steps. The following values are available: $\bar{x}_m = 34.77$, $s_m = 7.55$, $n_m = 65$, $\bar{x}_f = 33.29$, $s_f = 7.76$, $n_f = 35$.

13.2 We are interested in whether industrial firms or service firms growth faster over the last five years. Studies and theoretical considerations suggest that there should be a difference. Perform the test at the 5% significance level, considering all relevant steps. The following values are available: $\bar{x}_i = 6.35$, $s_i = 5.03$, $n_i = 34$, $\bar{x}_s = 7.49$, $s_s = 5.77$, $n_s = 66$.

13.3 Using Excel, rerun the test from Task 13.2 at the 5% significance level. Compare the results. Take into account any rounding errors.

13.4 We are interested in whether industrial firms or service firms spend more money on innovation. Studies and theoretical considerations indicate that industrial firms spend more. Perform the test at the 1% significance level, considering all relevant steps. The following values are available: $\bar{x}_i = 4.71$, $s_i = 3.47$, $n_i = 34$, $\bar{x}_s = 4.62$, $s_s = 3.38$, $n_s = 66$.

13.5 Using Excel, rerun the test from Task 13.4 at the 1% significance level.

13.6 We ask whether men and women have more job experience when they start an enterprise. Existing studies suggest that there is a difference. We perform the test at the 10% significance level, considering all relevant steps. The following values are available: $\bar{x}_m = 7.2$, $s_m = 3.42$, $n_m = 65$, $\bar{x}_f = 6.86$, $s_f = 3.67$, $n_f = 35$.

13.7 Using Excel, rerun the test from Task 13.6 at the 10% significance level.

13.8 We are interested in whether male or female founders spend more on marketing. Theoretical considerations lead us to assume that male founders spend more money on marketing than female founders. We conduct the test considering all relevant steps using Excel.

13.9 What values do we need to perform the test for a difference between group means?

13.10 If the test variable is metric but not normally distributed or ordinal, which test is used?

The Test for a Difference Between Means with Dependent Samples or Dependent Samples t-Test

14

14.1 Introduction to the Test for a Difference Between Means with Dependent Samples

In the test for a difference between group means with independent samples, we analyze whether two independent groups differ. We examine two different groups and then compare them. In the test for a difference between means with dependent samples, we examine the same group twice and ask whether a particular treatment has an effect on the group. For example, we might ask whether alcohol has an effect on a person's ability to drive, or whether a drug improves a person's health. We examine the person before and after the treatment, then compare the results and thus ask about the effect of the treatment. In a business context, for example, the method is used to evaluate the effect of an advertising campaign, e.g. to ask what influence an advertising campaign has on people's attitudes towards a particular product or company. However, one could also investigate how energy drinks influence the performance of athletes.

14.2 The Example: Training for Enterprise Founders in the Pre-founding Phase

The data that the test requires involves one of the few situations that our data set does not cover. Therefore, we need to introduce a new data set data_further_education.xlsx. Let's assume that the Young Entrepreneurs Association conducts further education for potential enterprise founders. These trainings are designed to help potential founders in the phase before starting their enterprise to correctly assess the market potential of their business. In order to check whether the training has an influence on the perception of the course

© Springer-Verlag GmbH Germany, part of Springer Nature 2023
F. Kronthaler, *Statistics Applied With Excel*,
https://doi.org/10.1007/978-3-662-64319-8_14

Table 14.1 Data set expected turnover before and after training

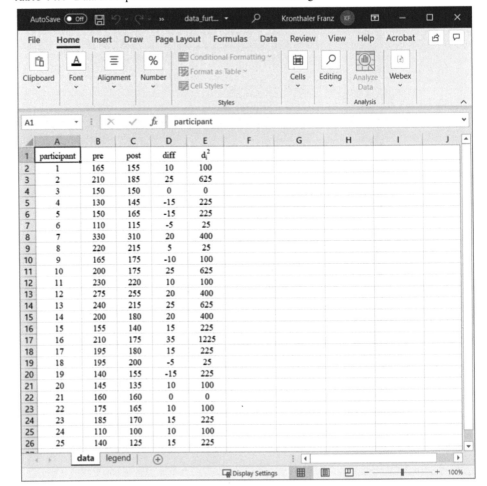

	A	B	C	D	E	F	G	H	I	J
1	participant	pre	post	diff	d_i^2					
2	1	165	155	10	100					
3	2	210	185	25	625					
4	3	150	150	0	0					
5	4	130	145	-15	225					
6	5	150	165	-15	225					
7	6	110	115	-5	25					
8	7	330	310	20	400					
9	8	220	215	5	25					
10	9	165	175	-10	100					
11	10	200	175	25	625					
12	11	230	220	10	100					
13	12	275	255	20	400					
14	13	240	215	25	625					
15	14	200	180	20	400					
16	15	155	140	15	225					
17	16	210	175	35	1225					
18	17	195	180	15	225					
19	18	195	200	-5	25					
20	19	140	155	-15	225					
21	20	145	135	10	100					
22	21	160	160	0	0					
23	22	175	165	10	100					
24	23	185	170	15	225					
25	24	110	100	10	100					
26	25	140	125	15	225					

participants, they are asked before and after the course about the expected turnover of their enterprise to be founded. The results are displayed in Table 14.1.

The legend to the data set can be found in Table 14.2.

We see that column B and C of the data set contains the turnover that the potential founders expect before after the training. Column D shows the difference. In column E, we have already calculated the squared difference, since we will need this information in a moment.

Table 14.2 Legend for the data set expected turnover before and after training

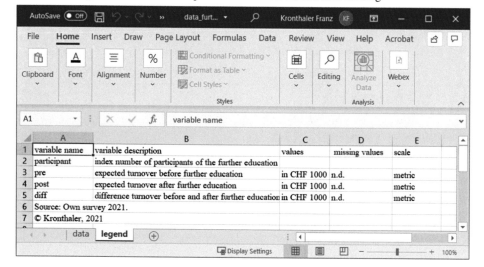

14.3 The Research Question and the Hypothesis in the Test: Does the Training have an Influence on the Market Potential Estimation?

The research question in our example is as follows: Does the training have an impact on business founders' assessment of the market potential of their proposed businesses? The null hypothesis and the alternative hypothesis are as follows:

H_0: The training has no influence on the market potential assessment of the founders.
H_A: The training has an influence on the market potential assessment of the founders.

or we can just write shorter:

H_0: $\mu_{pre} - \mu_{post} = 0$
H_A: $\mu_{pre} - \mu_{post} \neq 0$

Here are two more comments: First, we have formulated the null hypothesis and alternative hypothesis non-directional, since we have not made any statement regarding the direction of the influence. Second, we recognize that we compare the pre and post training means. μ_{pre} is the mean before the test and μ_{post} is the mean after the test in the population.

Let's furthermore determine the significance level. In this case, we want to test at the 10% significance level:

$$\alpha = 10\% \quad \text{or} \quad \alpha = 0.1$$

14.4   The Test Statistic

The test statistic is as follows:

$$t = \frac{\sum d_i}{\sqrt{\frac{n \sum d_i^2 - \left(\sum d_i\right)^2}{n-1}}}$$

with

d_i is the difference before and after the treatment and
n is the sample size or the number of people or objects observed.

We see that the test value is t-distributed and that we are using the test statistic to compare the difference between the pre- and post-treatment values. If the measure had no effect, the pre- and post-treatment values would be the same, so the difference would be zero, and thus the numerator of the test statistic would also be zero. It follows that the value of the test statistic would be zero as well and hence we would not reject the null hypothesis.

14.5   The Critical t-Value

Now we need to determine the critical t-values. We already know that these depend on whether we are testing two-sided or one-sided, how high the specified significance level is, and how many degrees of freedom we have. In our case, we have 25 observations. The number of degrees of freedom in the test is $df = n - 1$ i.e. we have $df = 25 - 1 = 24$ degrees of freedom. Thus, the critical t-values are -1.711 and 1.711.

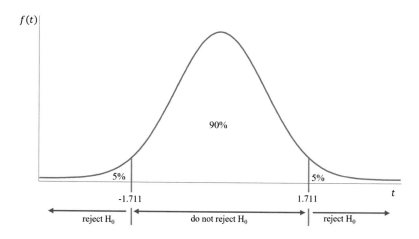

14.6   The t-Value and the Decision

Now all that is left is to compute the test statistic and make the test decision. For this, we insert the sum of the differences and the sum of the squared differences as well as the number of observed people in the test statistic. We obtain:

$$t = \frac{\sum d_i}{\sqrt{\dfrac{n\sum d_i^2 - \left(\sum d_i\right)^2}{n-1}}} = \frac{220}{\sqrt{\dfrac{25 \times 6'550 - 220^2}{25-1}}} = 3.17$$

If we compare the calculated t-value with the critical values, we see that we reject the null hypothesis and proceed with the alternative hypothesis. That is, the training has an effect on the market potential assessment of the enterprise founders.

14.7   The Effect Size

Before we come to the final interpretation, we calculate the effect size. For this we use the already known formula, with which we convert the t-value into the correlation coefficient r (see independent samples t-test):

$$r = \sqrt{\frac{t^2}{t^2 + df}}$$

with

t is the t-value of the test statistic and

df are the degrees of freedom.

The interpretation remains the same, of course:

- a $r < 0.3$ indicates a small effect,
- a r in the range between 0.3 to <0.5 means a medium effect,
- a r in the range of 0.5 to 1.0 indicates a large effect.

We insert the values of our example into the formula and see that we have a large effect:

$$r = \sqrt{\frac{t^2}{t^2 + df}} = \sqrt{\frac{3.17^2}{3.17^2 + 24}} = 0.54$$

Thus, we are able to evaluate our training. We reject the null hypothesis at the 10% significance level and thus assume that the training has an effect on how potential enterprise founders evaluate their market potential. The effect of the training is large. The mean value before the training was 183.4 (CHF 183'400), and after the training it was 174.6 (CHF 174'600). This means that the training causes the potential founders to assess the market potential after the training approximately 5% lower than before the training.

14.8   Calculating the Dependent Samples t-Test with Excel

For the calculation with Excel, the Data Analysis tool of Excel under the Data tab provides the following command: t-Test: Paired Two Sample for Means (see the figure).

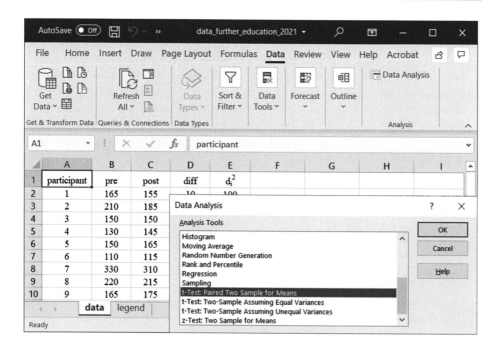

To perform the test for our example, we highlight the command, click on OK and get the following window:

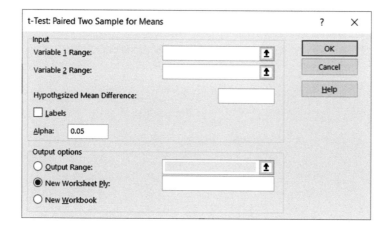

We now fill in the the required data. For variable *1* we enter the values before the training, and the values after the training for variable *2*. The hypothetical difference of the mean values is entered in the field Hypothesized Mean Difference. Here, we enter zero, because in our null hypothesis, we assume that the difference is zero. In Labels we put a check mark, this causes the variable name to be visible in the output. For Alpha, we enter our significance level. For example, if we test two-sided at the 10% level, we enter 0.1. In the Output Range, we specify the place in which we want the test result to appear. For example, we specify a new worksheet and enter a name for it. The input mask then looks as follows:

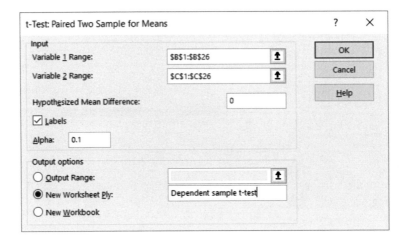

We click OK and get the results of our test in the worksheet defined.

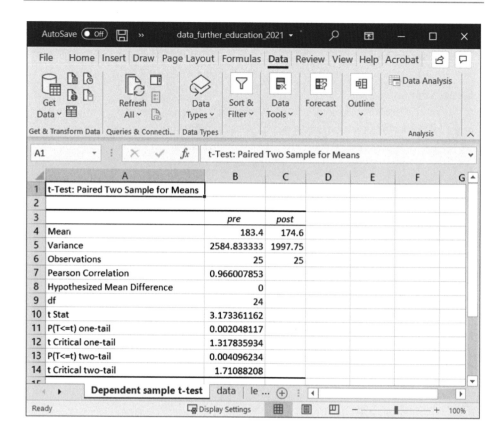

We should be familiar with most of the numbers. In row 4, we see the means before and after training. We had not calculated the variances of row 5. We do not need them either. In line 6 we see the number of observations. Again, we do not need the number of observations per variable but the number of pairwise comparisons, 25 in total. In row 7 we see the Pearson correlation coefficient. It informs about the correlation between the two variables, in our case it is close to 1, meaning that if the value of one variable gets smaller, then the value of the other variable gets smaller and vice versa. Line 8 displays the difference between the two means assumed in the null hypothesis. Line 9 gives the degrees of freedom back and line 10 obtains the value of the test statistic 3.173. We had calculated 3.17 by hand. Except for rounding errors, the values are identical to these calculated by hand. Now only the bottom four lines remain. In rows 12 and 14 we find the critical t-values for the one-sided and two-sided test. In rows 11 and 13 we find the probability for the one-sided and two-sided test of obtaining our calculated t-value, assuming that the null hypothesis is correct. We focus on row 13 since we tested two-sided. We see that the probability of obtaining the calculated t-value of 3.173 is 0.004 or 0.4%. Thus, the probability is much smaller than our specified significance level of 10% and we reject the null hypothesis. We also come to the same conclusion when we compare the calculated t-value with the critical t-value from line 14.

Freak Knowledge

If our data is metric but not normally distributed, then we can also use the test procedure described for a large sample. The test is relatively robust to a violation of the assumption of normally distributed data. For metric but not normally distributed data and a small sample, and for ordinal data, we have the Wilcoxon test for dependent samples.

14.9   Checkpoints

- *In the test for a difference between means in dependent samples, we address the question of whether a treatment has an influence on people or objects.*
- *In the test for a difference between means in dependent samples, we examine the same people or objects twice, once before and once after the treatment.*
- *The test assumes that the test variable is metric and normally distributed.*
- *For a small sample and metric but not normally distributed data and for ordinal data, we use the Wilcoxon test for dependent samples.*

14.10   Applications

14.1 We are interested in whether a training in accounting results in less time spent per month on accounting (data_accounting.xlsx). Perform the test by hand at the 10% significance level and interpret the result, considering all relevant steps.

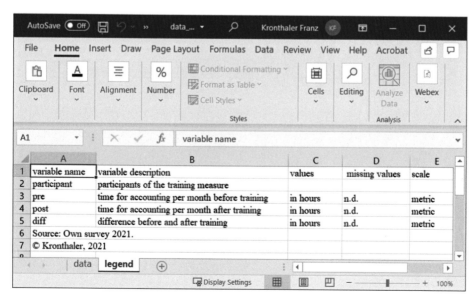

14.2 Perform the test from application 14.1 again at the 10% significance level using Excel and compare the results.

14.3 We are interested in whether drinking energy drinks affects the performance of athletes (data_drinks.xlsx). The study design involves the same athletes running as far as they can twice in one hour each time. Once they have the option to consume only water during the run, the other time they are allowed to consume only energy drinks. Run the test by hand at the 1% significance level and interpret the results, considering all relevant steps.

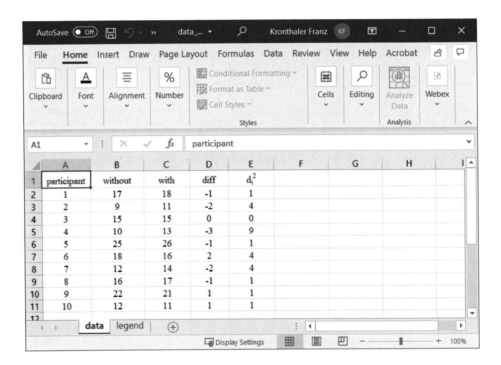

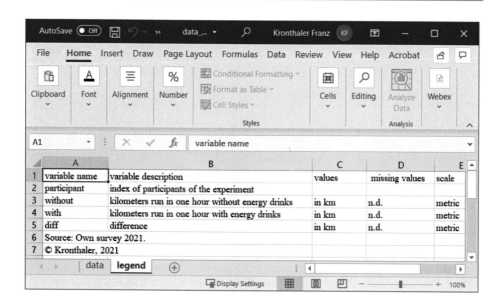

14.4 Perform the test from application 14.3 again at the 10% significance level using Excel
and compare the results.

The Analysis of Variance to Test for Group Differences When There Are More Than Two Groups

15

15.1 Introduction to the Analysis of Variance

In the test for a difference between group means with independent samples, we tested whether two groups differ with regard to the subject of interest, e.g. whether male founders and female founders are of the same age when they found their enterprise (cf. Chap. 13). But what do we do if we want to compare more than two groups at the same time. For example, the question might be: Do enterprise founders differ in their age by nationality? We then have more than two groups to compare. In such a case, the test for a difference between group means with independent samples is no longer sufficient, but instead analysis of variance can be used. Analysis of variance allows to compare more than two groups at the same time. The simplest form of an analysis of variance is the so-called one-way analysis of variance, with one test variable (founding age) and one grouping variable (nationality). The grouping variable is typically called factor. When we have two factors, e.g. nationality and gender, the analysis of variance is called two-way analysis of variance and so on.

In this chapter we deal only with the one-way analysis of variance (abbreviated one-way ANOVA). This gives a good introduction to ANOVA in general and it is relatively easy to extend the procedure with regard to more factors. In addition, we will learn about how to divide the variance observed in an explained part and an unexplained part, a very important concept in statistics.

Before we start now, a hint should be given here. For didactic reasons we introduce a new data set with data_workload.xlsx. The reason for this is didactical. In the first edition of the book, analysis of variance was not yet discussed and the use of the original data set would only lead to non-significant results, which is not expedient from a didactic point of view.

© Springer-Verlag GmbH Germany, part of Springer Nature 2023
F. Kronthaler, *Statistics Applied With Excel*,
https://doi.org/10.1007/978-3-662-64319-8_15

15.2 The Example: Do Enterprise Founders with Different Founding Motives Differ in the Amount of Time They Work?

Suppose we are interested in how much time enterprise founders work after founding their business. To analyze this, we survey the time they work in their founded enterprise per week. We also think that the amount of time worked per week depends on the motive for starting the business. The data collected could look as follows and are simulated, not real data (Table 15.1).

The legend to the data set is shown in Table 15.2.

If we sort the weekly workload by the motive for founding, the data can be displayed as in Table 15.3. In this table we see the weekly workload of the founders who founded out of unemployment (founding motive 1), who founded to implement an idea (founding motive 2), and who founded to achieve a higher income (founding motive 3).

Using this data, we can calculate four arithmetic means. We can calculate the average working time per founding motive. In addition, we can calculate the average working time for all observations.

$$\bar{y} = \frac{\sum y_{gk}}{n} = \frac{936}{18} = 52$$

$$\bar{y}_1 = \frac{\sum y_{1k}}{n_1} = \frac{262}{6} = 42$$

$$\bar{y}_2 = \frac{\sum y_{2k}}{n_2} = \frac{336}{6} = 56$$

$$\bar{y}_3 = \frac{\sum y_{3k}}{n_3} = \frac{348}{6} = 58$$

with

g	are the groups (the so-called factor levels),
k	are the observations per group respectively per factor level,
$\bar{y}$	is the global mean across all groups respectively factor levels,
$\bar{y}_g$	are the local means for the individual groups respectively factor levels.

Here, for the arithmetic mean $\bar{y}$ is used and not $\bar{x}$ as we learned in Chap. 3. This has to do with the fact that in the one-way analysis of variance we denote the factor with X and the test variable with Y. If we calculate the mean values for the test variable, these are described with $\bar{y}$ accordingly. In the end, it does not matter whether we label a variable with X, Y or Z.

Table 15.1 Data set workload and motive to found an enterprise

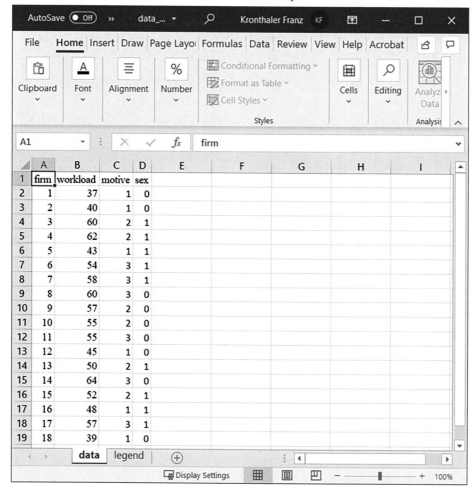

We see, all observed company founders work on average 52 h per week. The founders who started their business from unemployment work 42 h per week, the founders who want to implement an idea work 56 h per week and those who want to earn a higher income work 58 h per week.

In the data collected, we found differences with regard to the average weekly working hours. But do these differences in the sample also apply to the population? We analyze this question with the one-way ANOVA.

Table 15.2 Legend data set workload and motive to found an enterprise

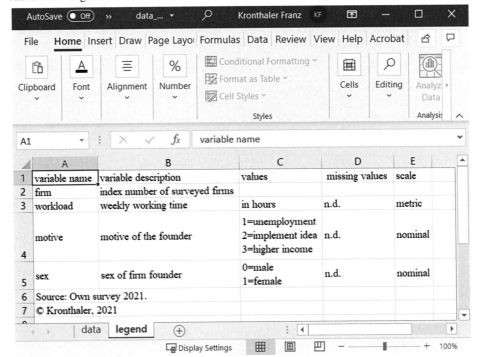

Table 15.3 Weekly workload in hours, sorted by enterprise foundation motive

Motive 1 = unemployment	Motive 2 = implement idea	Motive 3 = higher income
37	60	54
40	62	58
43	57	60
45	55	55
48	50	64
39	52	57

15.3 The Research Question and the Hypothesis of the Analysis of Variance

The research question for our example above is: Do enterprise founders with different founding motives differ in the amount of time they work? We can write the null hypothesis and the alternative hypothesis as follows:

H_0:   *The weekly working time of the enterprise founders differs not between the groups.*

H_A: *The weekly working time of the enterprise founders differs at least between two groups.*

or

$$H_0 : \mu_1 = \mu_2 = \mu_3$$
$$H_A : \mu_i \neq \mu_j \text{ for at least one pair of } ij$$

We see from the alternative hypothesis that we reject the null hypothesis if we find a group difference between two groups. As always, we specify the significance level, which should be here 5%:

$$\alpha = 5\% \ \text{ or } \ \alpha = 0.05$$

Now, before we turn to the test statistic, let's discuss the basic principle of analysis of variance. We will see that we can divide the variation of values around the mean into a into a part explained by the factor and a part not explained by the factor.

15.4   The Basic Idea of the Analysis of Variance

We have seen above that we can calculate a global mean value $\bar{y}$ for all observations and for each group a group mean value $\bar{y}_g$. To explain the principle of analysis of variance, it is best to plot both the observations y_{gk} as well as the mean values graphically (see Fig. 15.1).

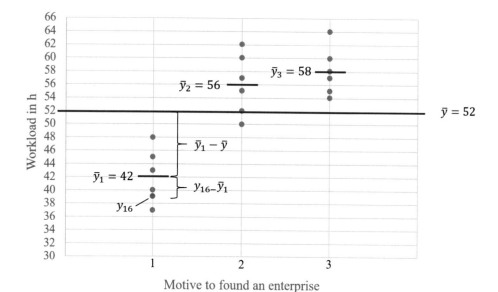

Fig. 15.1 Weekly workload of enterprise founders by motive

Table 15.4 Calculation of the total, explained and unexplained variance

enterprise	workload	motive	group means	variance total $(y_{gk}-y_{mw})^2$	variance explained $(y_{g,mw}-y_{mw})^2$	variance unexplained $(y_{gk}-y_{g,mw})^2$
1	37	1		225	100	25
2	40	1		144	100	4
5	43	1		81	100	1
12	45	1		49	100	9
16	48	1		16	100	36
18	39	1	42	169	100	9
3	60	2		64	16	16
4	62	2		100	16	36
9	57	2		25	16	1
10	55	2		9	16	1
13	50	2		4	16	36
15	52	2	56	0	16	16
6	54	3		4	36	16
7	58	3		36	36	0
8	60	3		64	36	4
11	55	3		9	36	9
14	64	3		144	36	36
17	57	3	58	25	36	1
		global mean	52	1168	912	256

We see in the figure the working hours of the enterprise founders, the mean for all enterprise founders and the group means. Let's focus on the red dot for group 1, which is observation y_{16}. The 1 stands for group 1 respectively founding motive 1 and the 6 for the sixth observation within the group. We see, this enterprise founder works significantly less than the average. The deviation is $y_{16} - \bar{y} = 39 - 52 = -13$ hours. It is the total deviation of this enterprise founder from the global mean. If we look at the point in more detail and include the group mean, we can divide the deviation into $\bar{y}_1 - \bar{y} = 42 - 52 = -10$ hours and $y_{16} - \bar{y}_1 = 39 - 42 = -3$ hours. We call the deviation of the group mean from the global mean the deviation explained by the factor level (founding motive 1), because this is the average group deviation from the average of all enterprise founders. The deviation of the observation from the group mean, on the other hand, is the unexplained deviation; we cannot explain this part of the deviation by the factor level. What we just did for this observation, can be done for all observations. If we do this and also square the deviations so that negative and positive deviations do not cancel each other out, then we have divided the total squared deviations (it is called sum of squares) into an explained part and an unexplained part (cf. Table 15.4).

15.5   The Test Statistic

The test statistic of the analysis of variance is derived from the explained and unexplained variance. The principle is simple, the higher the proportion of variance explained by the groups, the lower the unexplained proportion and the more likely the grouping variable contributes to explain the differences found. Specifically, the test statistic looks like this:

$$F = \frac{SS_{explained}/(G-1)}{SS_{unexplained}/(G \times (K-1))}$$

with

F	is the value of the test statistic,
$SS_{explained}$	is the explained sum of squares,
$SS_{unexplained}$	is the unexplained sum of squares,
$G - 1$	are the degrees of freedom of the numerator df_1,
$G \times (K-1)$	are the degrees of freedom of the denominator df_2.

We realize, the test statistic follows the F-distribution, that is, we can determine the critical value using the F-distribution table.

15.6   The Critical F-Value

From Chap. 10 we know that the critical value of the F-distribution is determined by two degrees of freedom, the degrees of freedom of the numerator df_1 and the denominator df_2, and by the chosen significance level of 10%, 5% or 1%. Thus, we take the specified significance level of 5%, calculate the two degrees of freedom with $df_1 = G - 1 = 3 - 1 = 2$ and $df_2 = G \times (K-1) = 3 \times (6-1) = 15$ and use these values to find the critical value in the F distribution table (Appendix E), it is 3.68. If our test statistic exceeds this value, then we reject the null hypothesis and proceed with the alternative hypothesis.

15.7   The F-Value and the Decision

To calculate the test statistic, we now only need to enter the values of the sum of squares explained and unexplained, as well as the degrees of freedom. By the way, if we divide the sums of squares by the respective degrees of freedom, we calculate the mean sums of squares. We will see this later when we calculate the ANOVA using Excel in this chapter.

The F-value is significantly higher than the critical value, which means that we reject the null hypothesis and proceed with the alternative hypothesis. We now know that there is

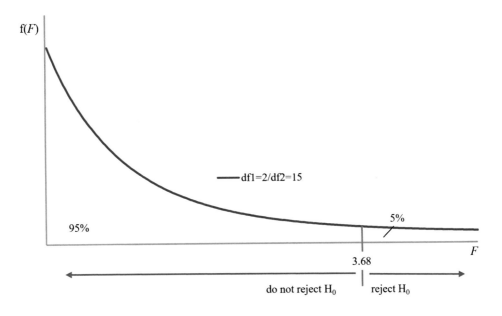

$$F = \frac{SS_{\text{explained}} / (G-1)}{SS_{\text{unexplained}} / (G \times (K-1))} = \frac{912/(3-1)}{256/(3 \times (6-1))} = \frac{456}{17.07} = 26.71$$

a high probability that there is a difference in weekly working hours between at least two groups in the population. However, we do not yet know between which groups there is a difference.

15.8   The Analysis of Variance an Omnibus Test and the Bonferroni Correction

As we have just shown, we use analysis of variance to test whether there is a group difference between at least two groups of the groups involved in the test procedure. We test all groups at once and do not differentiate between which groups there is a difference. The analysis of variance thus belongs to the so-called omnibus test procedures. In statistics, omnibus tests are test procedures that test for group differences, but do not specifically test between which groups differences exist.

In order to test between which groups there are differences, the so-called post hoc tests are used (they are called really that way). A widely used post hoc test in analysis of variance is the Bonferroni correction. We will discuss this method in more detail below.

Let's take a step back. We already know that a significant result indicates group differences in the population. However, we do not know between which groups there are differences. To determine this, we need to compare the groups pairwise. We can do this using the independent samples t-test (Chap. 13). However, there is still a small problem with this. If we compare all the groups pairwise, we need to do more than one test, how many tests depends on the number of groups. In our example we have three groups, that is we can compare group 1 with 2, group 1 with 3 and group 2 with 3. So we should do three tests.

In each of these tests performed, we allow an error equal to the specified significance level, and thus what is called error inflation occurs. Let's imagine that we make an error of 5% in the first test, an error of 5% again in the second, and an error of 5% again in the third. Then we have made a total error of 15%. But we do not want to increase the original significance level. Bonferroni has therefore suggested to divide the original significance level by the number of tests and then test against this new significance level:

$$\alpha^* = \frac{\alpha}{G \times (G-1)/2}$$

using

α^*	is the significance level at which each test is tested,
α	is the significance level at which the analysis of variance was tested,
G	is the number of groups respectively factor levels,
$G \times (G-1)/2$	is the number of tests to be performed.

Therefore, in our example, the significance level at which we should perform the pairwise comparisons is, $\alpha^* = 0.05/ (3 \times (3-1)/2) = 0.05/3 = 0.017$. If we perform the tests this way, we will find that we have a difference between groups 1 and 2 and 1 and 3. At this point, we can refresh the knowledge from Chap. 13 and do the tests by hand or we can wait until we introduce the calculation of the analysis of variance with Excel.

Freak Knowledge

As a possible post hoc test there is not only the Bonferroni correction available. Further post hoc tests would be, for example, the LSD procedure, Tukey's procedure, etc. Which method is used depends, among other things, on how strongly it is corrected for the error inflation. The Bonferroni correction is considered a very conservative procedure, i.e. it corrects very strongly for the error inflation.

15.9   The Effect Size

Now, before we finally interpret the results, let's see how the effect size is calculated in analysis of variance and determine it for our example.

As with the independent samples t-test and the dependent samples t-test, Bravais-Pearson's correlation coefficient can again be used. The squared correlation coefficient is calculated as follows:

$$r^2 = \frac{SS_{\text{explained}}}{SS_{\text{total}}}$$

In analysis of variance, this value is often referred to as Eta squared, denoted η^2. If we now take the root, we obtain the correlation coefficient as a measure of the effect size:

$$r = \sqrt{r^2} = \sqrt{\frac{SS_{\text{explained}}}{SS_{\text{total}}}}$$

We already know about the interpretation. A

- r less than < 0.3 indicates a small effect,
- a r in the range from 0.3 to < 0.5 indicates a medium effect, and
- a r in the range from 0.5 to 1.0 indicates a large effect.

For our example r is 0.88, i.e. we have a large effect.

$$r = \sqrt{\frac{SS_{\text{explained}}}{SS_{\text{total}}}} = \sqrt{\frac{912}{1168}} = 0.88$$

The result of our analysis can thus be presented as follows. The analysis of variance indicates a statistically significant result. We reject the null hypothesis on the significance level of 5% and assume that there are group differences. The effect size shows that we have a large effect or that the motive of starting a business has a strong effect on the weekly workload. If we compare the individual groups, we see that entrepreneurs with the motive of unemployment work significantly less than entrepreneurs with the motives of implementing an idea or achieving a higher income. The former work an average of 42 hours per week, the latter 56 and 58 hours per week. On average, the difference is 14 respectively 16 hours per week.

Freak Knowledge

Analysis of variance requires a metric and normally distributed test variable. In addition, the groups must be independent. If the group sizes vary widely, equal variances are required too. So what do we do when these assumptions are violated. If the test variable is metric but not normally distributed or ordinal, we have the Kruskal-Wallis test at our disposal. If the variances are not equal, then there is a solution for this with the Welch F-test. Whether the groups influence each other, i.e. are not independent of each other, should be clarified theoretically. There is no statistical procedure for this.

15.10 The Calculation of the Analysis of Variance with Excel

To calculate the analysis of variance with Excel, we have the Data Analysis tool in the Data tab with the command "Anova: Single Factor" (see the following figure). In order to perform the pairwise comparisons following the analysis of variance, we use the commands described in Chap. 13: t-Test: Two-Sample Assuming Equal Variances or t-Test: Two-Sample Assuming Unequal Variances.

It is best first to sort the data according to the groups and arrange them in columns. Then we click on the Data Analysis button in the Data tab and select the command "Anova: Single Factor". We then fill the command with the data, specify whether we have labels, specify the significance level and the place of the output.

After clicking OK we obtain the result and we can interpret it. If we keep an eye on the figures calculated in the example, we will see that the figure Excel gives out are similar. In order to draw the boxplot, we may go back to Chap. 5.

In line 1 we see that we performed a one-way analysis of variance. In rows 5 to 7 we see the number of observations, the mean values for the groups in the sample, as well as the group variances. Starting in row ten, the results of the analysis of variance are presented. Line twelve first shows the explained sum of squares (differences between groups), the degrees of freedom, and the explained mean sum of squares (the quotient of the explained sum of squares divided by the degrees of freedom). In row 13, we see the unexplained sum of squares, the degrees of freedom, and the unexplained mean sum of squares (again, the quotient of the sum of squares unexplained divided by the degrees of freedom). If we divide the mean sums of squares, we receive the F-value in row 12 column E. It is 26.72, as calculated in our example. Next to it are the P-value and the critical F-value. We took

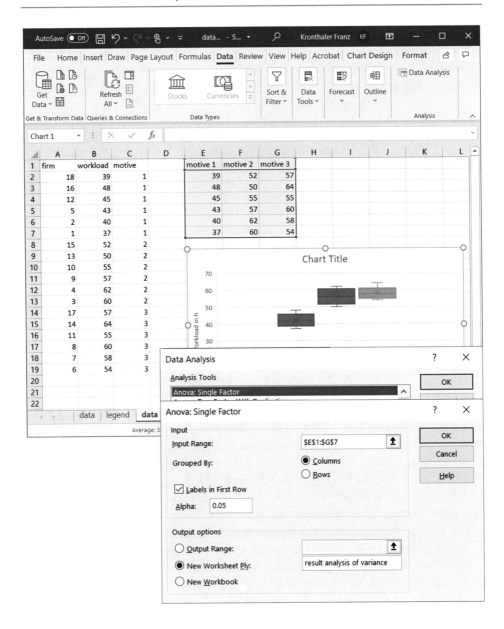

the critical F-value from the F-table with 3.68, here it is shown in more detail. We clearly exceed this with our F-value and reject the null hypothesis. The probability of obtaining such a high F-value if the null hypothesis would be correct is 0.0000113757 (P-value), i.e. it is much smaller than our specified significance level.

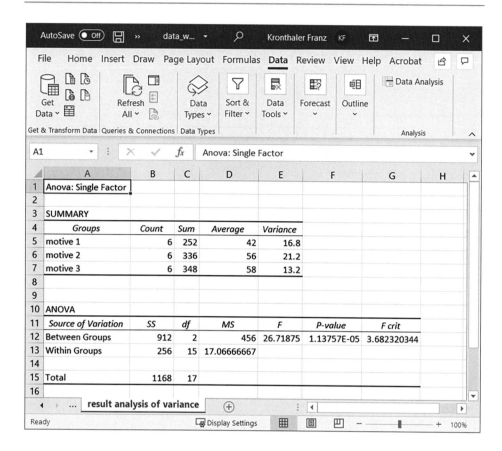

If we have a significant analysis of variance, we should check between which groups there is a difference. We do this using the independent samples t-test. The tests to be performed here are shown in the following figure (see also Chap. 13).

It can be seen that we reject the null hypothesis for the groups with motive 1 and motive 2 as well as motive 1 and motive 3, while we do not reject the null hypothesis for the group comparison between the groups with motive 2 and motive 3. We compare the yellow highlighted boxes for this purpose. The field in row 2 column C displays the significance level determined by the Bonferroni correction. We fall below this level at row 15 column B and column F. We do not fall below this level at row 29 column B. We look at the two-sided values since we are testing for group differences, these can be positive or negative.

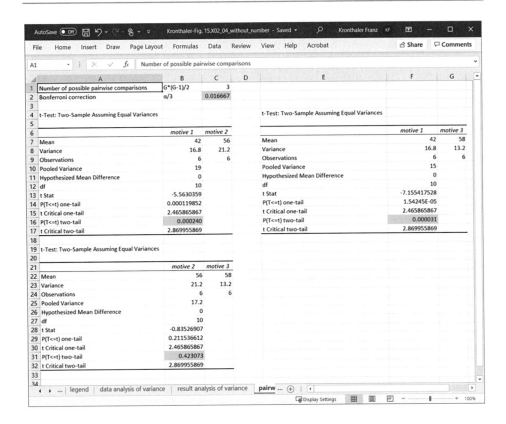

15.11 Checkpoints

- We use analysis of variance when we are comparing more than two groups at the same time.
- The analysis of variance is called ANOVA too.
- The analysis of variance is an omnibus test, we analyze here if there are group differences between all groups compared, not between which ones.
- Following the analysis of variance, a post hoc test should be performed to determine between which groups differences exist.
- When groups are compared pairwise, the so-called error inflation occurs, hence an error correction is necessary.
- If the test variable is metric but not normally distributed or if the test variable is ordinal, the Kruskal–Wallis test is used.

15.12 Applications

15.1 Use the first twelve observations of the data set data_workload.xlsx ($n = 12$) and recalculate the example in the textbook. Test at the 10% significance level whether there is a difference in the workload between the founders with different founding motives.

15.2 We are interested in whether there is a difference in enterprise growth between founders with different levels of education. Use the data set data_growth and test at a significance level of 5% with Excel the null hypothesis: The group means are the same.

15.3 We analyze the question of whether founders with different educational backgrounds spend different amounts of money on marketing. Use the data set data_growth and test at a significance level of 10% with Excel the null hypothesis: The group means are the same.

15.4 Use the data set data_workload.xlsx and analyze whether male and female founders differ in terms of their workload. Test at the 5% significance level using both the analysis of variance and the independent samples t-test. Compare the test results.

The Test for Correlation with Metric, Ordinal, and Nominal Data

In Chap. 6 we explored the question of the relationship between two variables, i.e. how and whether two variables move together. We analyzed this for metric and ordinal as well as nominal variables. For example, we examined a correlation between the variables marketing and innovation or between the variables expectation and self-assessment. We calculated the correlation coefficients based on the sample. With the test for correlation, we now check whether the correlation discovered in the sample also applies to the population. Only then is the sample result really relevant for recommendations. At the same time, the correlation coefficients also provide information about how strong the effect is. Thus, when we report the correlation coefficient, we do not need to calculate a measure of the effect size, the correlation coefficient is itself this measure. We report the correlation coefficient and whether it is significant. We then interpret the result accordingly.

In the following, the tests of correlation for metric, ordinal and nominal data are discussed. The procedure is the same as for the hypothesis tests already discussed. We first formulate the null hypothesis as well as the alternative hypothesis and specify the significance level. Then we determine the test distribution and the test statistic. After that we proceed with the rejection area and the critical values. As final steps, we calculate the value of the test statistic, decide and interpret the results.

16.1 The Test for a Correlation with Metric Data

16.1.1 The Test Situations for a Correlation with Metric Data

When we perform the test for correlation with metric data, we deal with three situations. First, we can assume that there is a correlation between two variables without specifying

© Springer-Verlag GmbH Germany, part of Springer Nature 2023
F. Kronthaler, *Statistics Applied With Excel*,
https://doi.org/10.1007/978-3-662-64319-8_16

the direction. Second, we can assume a positive correlation. Finally, it is possible that we think that there is a negative correlation between the variables.

In the first case, we test non-directional. The null hypothesis and the alternative hypothesis are then as follows:

H_0:   There is no relationship between variable X and variable Y.
H_A: There is a relationship between variable X and variable Y.

No relationship means that the correlation coefficient is zero. Accordingly, a relationship means that the correlation coefficient is not zero. Therefore, we can also write briefly as follows:

$$H_0: \quad r = 0$$

$$H_A: \quad r \neq 0$$

In the second case, we assume that there is a positive relationship in the population. The null and alternative hypotheses then are as follows:

H_0:   There is a negative or no relationship between variable X and variable Y.
H_A: There is a positive relationship between variable X and variable Y.

More briefly, we write:

$$H_0: \quad r \leq 0$$

$$H_A: \quad r > 0$$

If we reject the null hypothesis, we assume that there is a positive relationship between the variables in the population, the correlation coefficient is larger than zero.

In the third case, due to the assumption of a negative correlation, the null and alternative hypotheses are as follows:

H_0:   There is a positive or no relationship between variable X and variable Y.
H_A: There is a negative relationship between variable X and variable Y.

More briefly, we write:

$$H_0: \quad r \geq 0$$

$$H_A: \quad r < 0$$

If we reject the null hypothesis, we assume that the correlation coefficient is less than zero and there is a negative relationship in the population.

16.1.2 The Test Statistic and the Test Distribution

We perform the test for correlation with metric data using the correlation coefficient of Bravais-Pearson (compare Chap. 6).

$$r = \frac{\sum (x_i - \bar{x})(y_i - \bar{y})}{\sqrt{\sum (x_i - \bar{x})^2 \sum (y_i - \bar{y})^2}}$$

We calculate the correlation coefficient of Bravais-Pearson with the sample data and insert it into the following test statistic:

$$t = \frac{r \times \sqrt{n-2}}{\sqrt{1 - r^2}}$$

with

r is the correlation coefficient of Bravais-Pearson and
n is equal to the number of observations.

As we recognize, the test value is the t-value and our test distribution is the t-distribution with $df = n - 2$ degrees of freedom.

We possess now all the knowledge we need and can start with an example.

16.1.3 Example: Is There a Relationship Between Expenditure on Marketing and Expenditure on Innovation in Young Enterprises?

We would like to analyze the question whether there is a relationship between expenditure on marketing and expenditure on innovation in young enterprises. We read the literature and learn about the theory. In doing so, we find out that the budget of enterprises is limited and that money spend at one corner must be saved at somewhere else. We therefore formulate our null hypothesis and our alternative hypothesis as follows:

H_0:   *There is a positive or no relationship between the expenditure on marketing and the expenditure on innovation.*

H_A: There is a negative relationship between the expenditure on marketing and the
        expenditure on innovation.

$$H_0:\quad r \geq 0$$

$$H_A:\quad r < 0$$

Thus, we assume a negative correlation, i.e. enterprises, which spend a lot on marketing,
spend little on innovation and vice versa.

Let's test our hypothesis at the significance level of 5%.

$$\alpha = 5\% \quad \text{or} \quad \alpha = 0.05$$

In Chap. 6 we have already calculated the correlation coefficient on this issue for the
first six companies. The calculated value was $r = -0.91$. At this point it is certainly useful
to look at the example again (see Chap. 6).

To test whether there is indeed a correlation in the population, we put the value of the
correlation coefficient into our test statistic and calculate the t-value. We get

$$t = \frac{r \times \sqrt{n-2}}{\sqrt{1-r^2}} = \frac{-0.91 \times \sqrt{6-2}}{\sqrt{1-(-0.91)^2}} = -4.39$$

To proceed with the test decision we only need the critical t-value and then we can go
on. The test distribution is the t-distribution with $df = n - 2$ degrees of freedom, hence,
given 6 observations we have 4 degrees of freedom. With that and the significance level of
5% the critical value is -2.132 (compare Appendix C).

The calculated value of the test statistic falls below the critical value, i.e. we reject
the null hypothesis and continue working with the alternative hypothesis. Thus, in
the population, there is a negative correlation between expenditure for innovation and
expenditure for marketing. The magnitude of the correlation coefficient of -0.91 indicates
that there is a strong relationship or effect. So, enterprises which spend a lot on marketing
spend less on innovation and vice versa. If we like, we have discovered a specialization
strategy of young enterprises. Some try to succeed through marketing, others to position
themselves through innovation.

16.2   The Test for a Correlation with Ordinal Data

16.2.1 The Test Situations for a Correlation with Ordinal Data

The test situations when testing for a correlation with ordinal data are the same as with
metric data. The only difference is that we do not use Bravais-Pearson's correlation

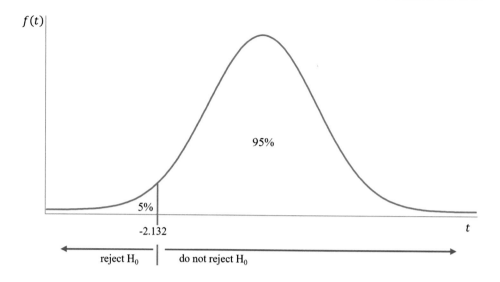

coefficient, but Spearman's correlation coefficient (compare Chap. 6):

$$r_{Sp} = \frac{\sum \left(R_{x_i} - \overline{R}_x\right)\left(R_{y_i} - \overline{R}_y\right)}{\sqrt{\sum \left(R_{x_i} - \overline{R}_x\right)^2 \sum \left(R_{y_i} - \overline{R}_y\right)^2}}$$

Like the Bravais-Pearson correlation coefficient, Spearman's correlation coefficient can be both positive or negative. Thus, we can test non-directional, for a positive relationship, and for a negative relationship. Let's write down the null hypothesis and the alternative hypothesis in the short form for these three situations.
Situation 1:

$$H_0: \quad r_{Sp} = 0$$
$$H_A: \quad r_{Sp} \neq 0$$

Situation 2:

$$H_0: \quad r_{Sp} \leq 0$$
$$H_A: \quad r_{Sp} > 0$$

Situation 3:

$$H_0: \quad r_{Sp} \geq 0$$
$$H_A: \quad r_{Sp} < 0$$

16.2.2 The Test Statistic and the Test Distribution

The test statistic used to perform the test is t-distributed with $df = n - 2$ degrees of freedom and is as follows:

$$t = \frac{r_{Sp}}{\sqrt{\dfrac{1 - r_{Sp}^2}{n - 2}}}$$

where

r_{Sp} is the correlation coefficient of Spearman and
n is the number of observations.

With this, we can already perform the test and start with an example.

16.2.3 Example: Is There a Relationship Between Self-assessment and Expectation Regarding the Economic Development of an Enterprise?

In Chap. 6 we analyzed whether there is a relationship between the variable expectation and the variable self-assessment. The assumption behind this was that the self-assessment in terms of industry job experience and expectation in terms of future development of the enterprise may be correlated. In Chap. 6 we calculated a positive correlation for the first six companies amounting to $r_{Sp} = 0.58$. Now we want to test whether this relationship also holds for the population. The null hypothesis and the alternative hypothesis are as follows:

$$H_0: \quad r_{Sp} \leq 0$$
$$H_A: \quad r_{Sp} > 0$$

We would like to test at the 10% significance level.

$$\alpha = 10\% \quad \text{or} \quad \alpha = 0.1$$

To find out if the correlation holds for the populations, we insert the value of the correlation coefficient and the number of observations into our test statistic and calculate

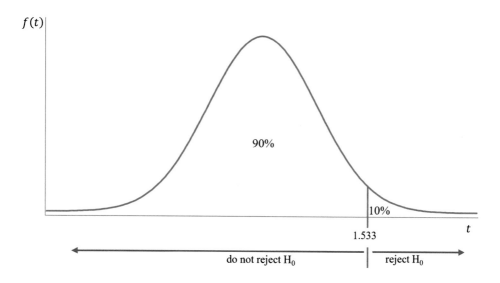

the t-value. We obtain:

$$t = \frac{r_{Sp}}{\sqrt{\dfrac{1 - r_{Sp}^2}{n - 2}}} = \frac{0.58}{\sqrt{\dfrac{1 - 0.58^2}{6 - 2}}} = 1.424$$

We now still need to compare this value with our critical value, which we obtain from the t-table. We reject the null hypothesis if we exceed a value of 1.533 (compare Appendix C).

If we compare our calculated test value of 1.424 with the critical value, we see that we do not reject the null hypothesis. This means that there is no positive relationship in the population. So, there is no positive relationship between the expectation about the future development of the enterprise and the self-assessment regarding the industry job experience. Here a small note on the effect size. The calculated correlation coefficient of 0.58 indicates a large effect (compare Sect. 11.3). However, the sample is with $n = 6$ relatively small, and is not sufficient large to reject the null hypothesis at this magnitude of the correlation coefficient. However, it would probably be worth to increase the sample slightly.

16.3   The Test for Correlation with Nominal Data

16.3.1 The Test Situations When Testing for a Correlation with Nominal Data

When we test for correlation between nominal variables, we have two situations. Either we have nominal variables with two characteristics each, or we have nominal variables with more than two characteristics. In both cases, the correlation coefficient says nothing about the direction of the relationship, that is, we cannot test for a positive or a negative relationship. We can only test for a relationship. For this reason, both test procedures are often called tests of independence. If the variables are not correlated, then they are independent of each other.

If we have two nominal variables with two characteristics each, then the null hypothesis and alternative hypothesis is as follows:

H_0:   *There is no correlation between variable X and variable Y.*
H_A: *There is a correlation between variable X and variable Y.*

Remember, the relevant correlation coefficient is the phi coefficient r_Φ (see Chap. 6). r_Φ can assume positive and negative values, whereby the sign makes no statement about the direction of the correlation. We can therefore write briefly as follows:

$$H_0: \quad r_\Phi = 0$$

$$H_A: \quad r_\Phi \neq 0$$

For nominal variables with more than two expressions, the contingency coefficient C comes into play. The null hypothesis and alternative hypothesis are the same.

H_0:   *There is no relationship between variable X and variable Y.*
H_A: *There is a correlation between variable X and variable Y.*

However, since the contingency coefficient only takes values greater than and equal to zero, we briefly write as follows:

$$H_0: \quad C = 0$$

$$H_A: \quad C > 0$$

In both cases, if the null hypothesis is rejected, we assume that there is a relationship between the variables, they are not independent of each other.

16.3.2 The Test of Independence for Nominal Variables with Two Characteristics

For nominal variables with two characteristics, we calculated the phi coefficient:

$$r_\Phi = \frac{a \times d - b \times c}{\sqrt{S_1 \times S_2 \times S_3 \times S_4}}$$

where
a, b, c, d are the fields of a 2×2 matrix and
S_1, S_2, S_3, S_4 are the row sums and column sums.

		Variable Y		
		0	1	
Variable X	0	a	b	S_1
	1	c	d	S_2
		S_3	S_4	

The value r_Φ is also our test statistic. The squared value of r_Φ multiplied by the number of observations is $\chi 2$-distributed with $df = 1$ degrees of freedom:

$$\chi^2 = n \times r_\Phi^2$$

Thus we know our test statistic, which is $\chi 2$-distributed and we can perform the test of independence.

In Chap. 6 we analyzed the question whether there is a relationship between the sex of the founders and the sector in which they found. In doing so, we used the first ten enterprises and calculated, using the following 2×2 *matrix* a r_Φ in the magnitude of -0.50.

		Sex		
		0 = male	1 = female	
Sector	0 = industry	2	0	2
	1 = service	3	5	8
		5	5	

Hence we have discovered a medium correlation between sex and the sector in which the enterprise was founded. If we look at the matrix, we see that female founders are more likely to start service enterprises, while male founders are more likely to start industrial companies. But does the correlation hold for the population?

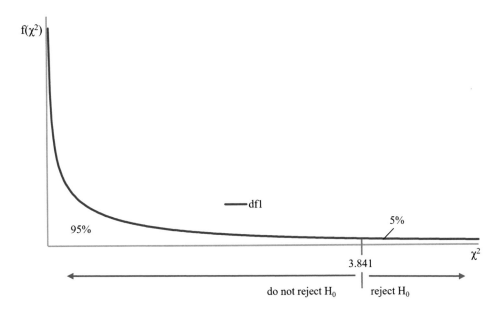

For this, we conduct the test where the null and alternative hypotheses are as follows:

H_0:   *There is no relationship between the sex of a founder and the sector in which the enterprise is founded.*

H_A: *There is a relationship between the sex of a founder and the sector in which the enterprise is founded.*

$$H_0: \quad r_\Phi = 0$$

$$H_A: \quad r_\Phi \neq 0$$

We test this at the 5% significance level:

$$\alpha = 5\,\% \quad \text{or} \quad \alpha = 0.05$$

With $r_\Phi = -0.50$ we calculate the value of the test statistic. This is the $\chi 2$ value:

$$\chi^2 = n \times r_\Phi^2 = 10 \times (-0.50)^2 = 2.50 \; with \; df = 1.$$

We now compare the value of the test statistic with the critical value out of the $\chi 2$-distribution table (see Appendix D). We see that we reject the null hypothesis if we exceed a value of 3.841.

If we compare our calculated test value with the critical value, the result is that we do not reject the null hypothesis, i.e., based on our sample finding, we have no relationship between the two variables in the population. Again, despite a relatively large effect of $r_\Phi = 0.50$ we assume no correlation in the population. The reason is, as above for the test with ordinal data, again the small sample size. In addition, this test only provides reliable results if each cell contains at least 5 objects. For didactic reasons, we worked with a small number of observations. Therefore, all cells except one contain less than 5 observations. This means that we have to increase the number of observations in order to perform the test in a reliable way.

16.3.3 The Test of Independence for Nominal Variables with More Than Two Characteristics

For nominal variables with more than two characteristics, the contingency coefficient is used (see Chap. 6):

$$C = \sqrt{\frac{U}{U+n}}$$

with
U is the sum of the deviations between the observed and theoretically expected values:

$$U = \sum \sum \frac{(f_{jk} - e_{jk})^2}{e_{jk}}$$

where

f_{jk} are the observed frequencies and
e_{jk} *are the theoretically expected frequencies.*

For example, if we have a nominal variable with two characteristics and one with three characteristics, then the matrix of the observed frequencies is as follows.

Rows	Columns		
	1	2	3
1	f_{11}	f_{12}	f_{13}
2	f_{21}	f_{22}	f_{23}

The theoretical expected frequencies are calculated the following way:

Rows	Columns			
	1	2	3	Row sums
1	$e_{11} = \frac{f_1 \times f_1}{n}$	$e_{12} = \frac{f_1 \times f_2}{n}$	$e_{13} = \frac{f_1 \times f_3}{n}$	f_1
2	$e_{21} = \frac{f_2 \times f_1}{n}$	$e_{22} = \frac{f_2 \times f_2}{n}$	$e_{23} = \frac{f_2 \times f_3}{n}$	f_2
Column sums	f_1	f_2	f_2	

From the observed and the theoretically expected frequencies, we then calculated the deviation sum U. The deviation sum U is our test statistic. It is χ^2-distributed with $df = (j - 1) \times (k - 1)$ degrees of freedom, where j is the number of rows and k is the number of columns.

With this we know our test distribution, it is the χ^2-distribution and we know our test statistic, which is U and χ^2-distributed. Hence, we are able to perform the test for independence for nominal variables with more than two characteristics.

In our example from Chap. 6 we analyzed whether there is a relationship between the motive for founding an enterprise and the sector in which the business was founded. In doing so, we observed the following cell frequencies for our 100 firms:

Sector	Motive			
	Unemployment	Implement idea	Higher income	Row sums
Industry	8	15	11	34
Service	9	39	18	66
Column sums	17	54	29	

and calculated the following theoretical frequencies:

Sector	Motive			
	Unemployment	Implement idea	Higher income	Row sums
Industry	$e_{11} = \frac{34 \times 17}{100} = 5.78$	$e_{12} = \frac{34 \times 54}{100} = 18.36$	$e_{13} = \frac{34 \times 29}{100} = 9.86$	34
Service	$e_{21} = \frac{66 \times 17}{100} = 11.22$	$e_{22} = \frac{66 \times 54}{100} = 35.64$	$e_{23} = \frac{66 \times 29}{100} = 19.14$	66
Column sums	17	54	29	

Out of this U was calculated with a value of 2.42 and the contingency coefficient C with 0.15. Given the small value of the contingency coefficient, we had claimed that there is no relationship between the two variables. We can now check whether the variables are indeed independent in the population.

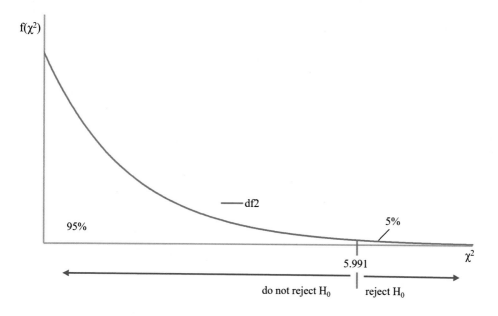

The null and alternative hypotheses are:

H_0:   *There is no relationship between the motive to start an enterprise and the sector in which it is founded.*

H_A: *There is a relationship between the motive to start an enterprise and the sector in which it is founded.*

$$H_0: \quad C = 0$$

$$H_A: \quad C > 0$$

We want to test this at the 5% significance level.

$$\alpha = 5\% \quad \text{or} \quad \alpha = 0.05$$

Our test statistic is, as stated above, U with a value of 2.42. It is $\chi 2$-distributed with $df = (j - 1) \times (k - 1) = (2 - 1) \times (3 - 1) = 2$ degrees of freedom. Thus the critical value comes from the $\chi 2$ distribution table and is 5.991 (compare Appendix D).

If we compare the test value of 2.42 with the critical value, we find that we do not reject the null hypothesis. Based on the sample finding, we can make the statement that there is no relationship between the motive to start an enterprise and the sector in which it is founded. This result was expected because the contingency coefficient, which we can also take as a measure of the effect size, is rather small in the sample.

Again, it is important to make sure that each cell is occupied by at least 5 objects. Otherwise, the test can lead to incorrect results.

16.4   Calculating Correlation Tests with Excel

In Excel there is no command available to calculate the tests of correlation for metric and ordinal data. We calculate the correlation coefficients as shown in Chap. 6 and then test using the procedure described above.

For nominal variables, we have the CHIQU.TEST function available under the Formulas tab and the Insert Function button. The CHIQU.TEST function gives back the probability of no correlation between the variables, when we put in the observed frequencies and the theoretically expected frequencies. In other words, we first have to calculate the observed and the theoretically expected frequencies.

We then call the CHIQU.TEST function and fill the input mask with the values needed.

We click OK and get the probability of 0.298 or 29.8%. This is the probability of obtaining a test value equal to 2.42 and higher (see example above), assuming that the null hypothesis is correct. So, as above, we do not reject the null hypothesis based on the finding. We would only reject if this value falls below our specified significance level.

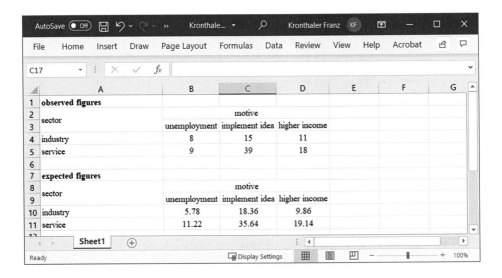

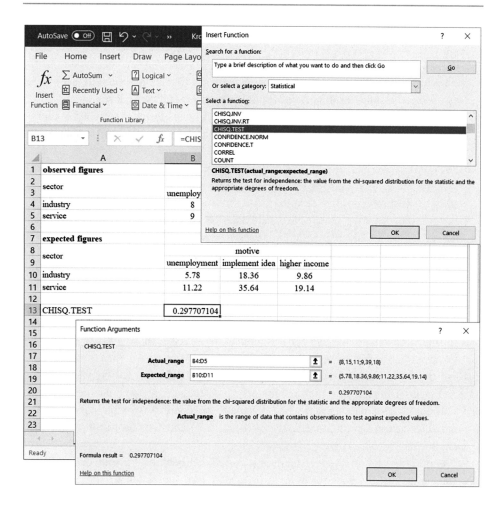

16.5 Checkpoints

- When testing for correlation, we use the correlation coefficients relevant to the data, such as Bravais-Pearson's correlation coefficient for metric data.
- For metric and ordinal variables, we can test non-directional and directional.
- For metric and ordinal variables, the test statistic is t-distributed.
- For nominal variables, we test whether there is a relationship. These test procedures are often called test for independence.
- For nominal variables, the test variable is $\chi 2$-distributed, and the test procedures are often called $\chi 2$-independence tests.

16.6   Applications

16.1 For our data set data_growth.xlsx, calculate the Bravais-Pearson correlation coefficient for the variables growth and marketing for the first eight observations ($n = 8$) by hand and test the result non-directional at the 5% significance level.

16.2 For our data set data_growth.xlsx, calculate the Bravais-Pearson correlation coefficient for the variables growth and marketing using Excel (n = 100) and test the result non-directional at the 5% significance level.

16.3 For our data set data_growth.xlsx, calculate the Bravais-Pearson correlation coefficient for the variables growth and experience for the first eight observations ($n = 8$) by hand. Test for a positive relationship at the 1% significance level.

16.4 For our data set data_growth.xlsx, calculate the Bravais-Pearson correlation coefficient for the variables growth and experience using Excel ($n = 100$). Test for a positive relationship at the 1% significance level.

16.5 For our data set data_growth.xlsx, calculate Spearman's rank correlation coefficient for the variables growth and self-assessment for the first eight observations ($n = 8$) by hand and test for a positive relationship at the 10% significance level.

16.6 For our data set data_growth.xlsx, calculate Spearman's rank correlation coefficient for the variables growth and self-assessment using Excel ($n = 100$) and test for a positive relationship at the 10% significance level.

16.7 For our dataset data_growth.xlsx, calculate the phi coefficient for the variables sex and sector using Excel ($n = 100$) and test the result at the 5% significance level.

16.8 For our dataset data_growth.xlsx, calculate the contingency coefficient for the variables sex and motive using Excel ($n = 100$) and test the result at the 10% significance level.

16.9 We are interested in whether there is a correlation between three stocks and look for the historical values for VW, Daimler and SAP (see Chap. 6). Test at the 5% significance level between which stocks there is a significant relationship.

Date	VW	Daimler	SAP
01.04.2020	99.27	25.74	97.40
31.03.2020	105.48	27.14	101.48
30.03.2020	105.50	27.21	101.70
27.03.2020	106.50	27.32	100.78
26.03.2020	112.90	29.21	103.88
25.03.2020	115.50	30.25	100.00

More Tests for Nominal Variables

By now we have learned about a lot of test procedures. We are still missing an important strand, the test procedures for nominal data. We will now take a look at these. The test procedures for nominal data are often called χ^2-tests, because the test distribution is usually the χ^2-distribution. We have already learned about one of the procedures for nominal data, the χ^2-test of independence. We will add the tests for nominal data with one sample, with two independent samples, and with two dependent samples.

17.1 The χ^2-Test with One Sample: Does the Share of Female Founders Correspond to the Gender Share in Society?

In the χ^2-test with one sample, we examine whether the observed frequencies in a sample match with the frequencies we have due to our prior knowledge. Looking at our data set, for example, we might ask whether or not the proportion of female founders is consistent with the proportion in society, i.e., whether significantly fewer or significantly more women start a business. However, we could also analyze whether the proportion of female founders in the sample matches the proportion of female founders in the population. In the latter case, we would check our sample for representativeness.

Let's go through the procedure using the first question. We ask whether more or fewer women start a business than men. The null hypothesis and the alternative hypothesis are then as follows:

H_0: *The frequency of firms founded by men and women is equal to the gender share in the society.*

H_A: *The frequency of firms founded by men and women is not equal to the gender share in the society.*

© Springer-Verlag GmbH Germany, part of Springer Nature 2023
F. Kronthaler, *Statistics Applied With Excel*,
https://doi.org/10.1007/978-3-662-64319-8_17

We know that there are roughly equal numbers of women and men in society, so we can write more briefly as follows:

$$H_0: \quad \pi_1 = \pi_2 = 0.5$$

$$H_A: \quad \pi_1 \neq \pi_2 \neq 0.5$$

where

π_1 and π_2 are the proportions of men and women in the population.

We now want to test this at the 10% significance level.

$$\alpha = 10\% \quad \text{or} \quad \alpha = 0.1$$

The test statistic is built from the deviations between the observed frequencies and the expected frequencies, i.e. we examine how the observed values differ from the theoretically expected values:

$$U = \sum \frac{(f_i - e_i)^2}{e_i}$$

where

f_i are the observed values and
e_i are the theoretically expected values.

The test statistic is χ^2-distributed with $df = c - 1$ degrees of freedom. c is the number of categories that can occur for the variable.

Now we need the observed values and the theoretically expected values. We observed a total of 35 female founders and 65 male founders in our sample of 100 firms. We would expect $100 \times 0.5 = 50$ female founders and 50 male founders, based on our gender share in the society. We insert these numbers into our formula:

$$U = \sum \frac{(f_i - e_i)^2}{e_i} = \frac{(65 - 50)^2}{50} + \frac{(35 - 50)^2}{50} = 9.0$$

We have $df = c - 1 = 2 - 1 = 1$ degrees of freedom. There are two possible characteristics, male and female, which means we have two categories.

We now only need to determine our critical χ^2-value and can then make the test decision. According to the χ^2-distribution table, the critical χ^2-value is 2.706 with one degree of freedom and a significance level of 10% (compare Appendix D).

Let's compare the calculated value of the test statistic U of 9.0 with the critical value of 2.706. We reject the null hypothesis. It follows, the frequency of firms founded by men and women does not correspond to the gender proportion in the population. Men are more

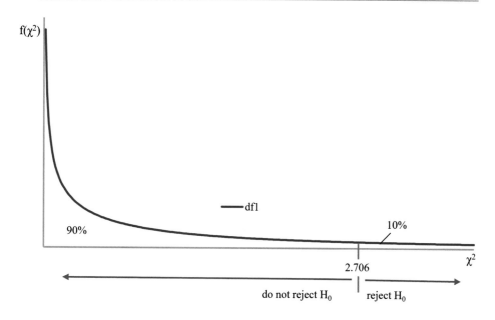

likely to start a business than women. The magnitude of the effect can be read from the figures. In absolute terms, we have 35 female founders and 65 male founders in the sample, i.e. 35% of the enterprises are founded by women and 65% by men.

Freak Knowledge

Measures used to calculate the effect size in χ^2-tests are, for example, the coefficient Phi or Cramer's V.

17.2   The χ^2-Test with Two Independent Samples: Are Start-Up Motives the Same for Service and Industrial Firms?

We use the χ^2-test with two independent samples to analyze whether two groups differ with respect to the frequency proportions of a nominal variable. We compare whether the respective frequencies in each group differ from the theoretically expected frequencies.

The best way to explain this is to use the example used in Chap. 16 in the test for independence. Here we had investigated whether there was a correlation between the motive for founding and the sector in which the enterprise is founded. We had observed the following cell frequencies for our 100 enterprises:

Sector	Motive			Row sums
	Unemployment	Implement idea	Higher income	
Industry	8	15	11	34
Service	9	39	18	66
Column sums	17	54	29	

The cell frequencies can be read as follows: In the industrial sector, 8 out of 34 firms (23.5 %) are founded out of unemployment, 15 out of 34 (44.1 %) of the founders want to implement an idea and 11 out of 34 (32.4 %) want to achieve a higher income. For service enterprises, the ratios are 9 to 66 (13.6%), 39 to 66 (59.1%) and 18 to 66 (27.3%).

The test is designed to determine whether the two groups differ in terms of the proportions of founding motives. We test the following null and alternative hypothesis:

H_0: *The proportions of the motives to found an enterprise are the same for industrial and service enterprises.*

H_A: *The proportions of the motives to found an enterprise are not the same for industrial and service enterprises.*

The procedure of the test is the same as the procedure of the test for independence. Again, we compare the observed frequencies with the theoretically expected frequencies. The test statistic is identical, it is again the deviation sum U:

$$U = \sum \sum \frac{\left(f_{jk} - e_{jk}\right)^2}{e_{jk}}$$

where U is χ^2-distributed with $df = (j - 1) \times (k - 1)$ degrees of freedom.

Thus the test is the same as discussed in Chap. 16. The difference lies only in the formulation of the hypotheses; one time we test for independence, the other time for difference between two groups. We take the calculation of the difference sum from Chap. 16. The value is 2.42 with 2 degrees of freedom. Thus the test decision is the same as above. We do not reject the null hypothesis.

17.3   The χ^2-Test with Two Dependent Samples: Is My Advertising Campaign Effective?

Sometimes we wonder whether an action changes the behavior of people or objects, for example, whether an advertising campaign carried out by our company has an effect on whether people buy our product or not. The null and alternative hypotheses would be:

H_0:   *The advertising campaign has no effect on whether people buy our product.*
H_A: *The advertising campaign has an influence on whether people buy our product.*

We want to test these hypotheses at the 1% significance level:

$$\alpha = 1\% \quad \text{or} \quad \alpha = 0.01$$

To conduct the test, let's suppose we draw a random sample of $n = 300$ people and ask them before our advertising campaign whether they are buyers or non-buyers of our product. After the campaign, we survey the same individuals again. Following this, we can construct the following 2×2 matrix:

		Survey after campaign	
		Buyers	Non-buyers
Survey before campaign	Buyers	$f_a = 80$	$f_b = 50$
	Non-buyers	$f_c = 100$	$f_d = 70$

Where the rows represent the situation before our campaign and the columns represent the situation after our campaign. f_a to f_d are the observed frequencies in cells a to d.

We can read the matrix as follows: Before the campaign, we had 80 + 50 = 130 buyers and 100 + 70 = 170 non-buyers in the sample. After the campaign, we have 80 + 100 = 180 buyers and 50 + 70 = 120 non-buyers. Looking more closely at the matrix, we find that of our original 130 buyers, 50 have switched to non-buyers and of our original 170 non-buyers, 100 have switched to buyers. In total, we have 150 switches in the sample.

The test procedure that now compares whether this switching occurs by chance, or whether our advertising campaign has an effect on it, is the McNemar χ^2-test. The test considers only the switchers, i.e. the cells on the top right (b) and bottom left (c) of the matrix. It assumes that one half of the switchers should be buyers to non-buyers and the other half non-buyers to buyers. To generalize this a bit, we can also say that half of the switchers change from 0 to 1 and the other half change from 1 to 0.

Knowing this, we can calculate the theoretical expected values for cells b and c:

$$e_b = e_c = \frac{f_b + f_c}{2}.$$

The theoretical expected frequency for cells b and c is thus:

$$e_b = e_c = \frac{f_b + f_c}{2} = \frac{50 + 100}{2} = 75.$$

Using the observed values and the theoretically expected values, we can in turn compute the sum of deviations, which also serves as the test statistic. It is χ^2-distributed with

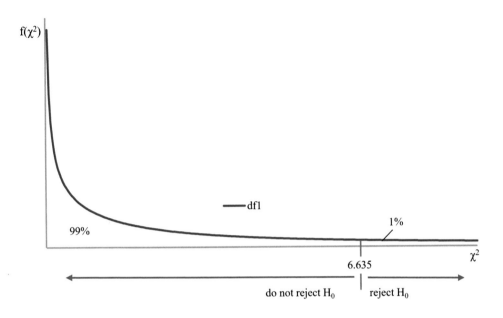

$df = 1$ degree of freedom:

$$U = \sum \frac{(f_i - e_i)^2}{e_i} = \frac{(f_b - e_b)^2}{e_b} + \frac{(f_c - e_c)^2}{e_c} = \frac{(50 - 75)^2}{75} + \frac{(100 - 75)^2}{75} = 16.7.$$

Now we just need to determine the critical χ^2-value. We take it from the χ^2-distribution table, it is 6.635 (compare Appendix D).

If we compare the calculated value with the critical value, we see that we reject the null hypothesis. So we can assume that our advertising campaign is working. We have significantly more buyers after the campaign, the change from non-buyers to buyers is significantly higher than the change from buyers to non-buyers.

Finally, we want to make an important note. For χ^2-test procedures to be effective, each cell should include at least 5 observations. If this is not the case, the test results are not trustworthy.

Freak Knowledge
In each of the χ^2-test procedures discussed here, we tested in an undirected manner. For χ^2-tests, this is the typical case. If there is a 2×2 matrix, we can also test directionally. In such a case, we follow the logic of the undirected test procedures and double the significance level.

17.4 Calculating the Tests with Excel

In Excel, we only have the χ^2-test for independence. We can also use this to perform the χ^2-test with two independent samples. The procedure is described in Chap. 16.

17.5 Checkpoints

- Test procedures for nominal data are based on frequency analysis.
- Test procedures for nominal data are often called just χ^2-tests.
- For χ^2-test procedures cell frequencies should be at least five observations per cell.

17.6 Applications

17.1 We want to know whether there are differences with respect to the motives to start an enterprise. To do so, we consider the following null hypothesis: all start-up motives occur with equal frequencies. Calculate the corresponding test procedure, test at the 5% significance level and interpret the result.

17.2 We are interested in whether there are differences between men and women with regard to the importance of motives to found an enterprise. Calculate the corresponding test procedure, test at the 5% significance level and interpret the result.

17.3 We are conducting an educational campaign in our company considering nutrition in the workplace. Our goal is to promote a health-conscious diet in the workplace. We are randomly observing 200 employees and asking them about their diet at work before and after the campaign. Using these results, we can construct the following 2×2 matrix. Calculate the appropriate test procedure, test at the 10% significance level and interpret the result.

		Survey after education	
		Health conscious	Non-health conscious
Survey before education	Health conscious	60	40
	Non-health conscious	70	30

Summary Part IV: Overview of Test Procedures 18

At the end of the fourth part of the book it is useful to summarize the discussed test procedures in a table. This table, in addition to Fig. 11.5, can serve as assistance in choosing a test procedure. It should be noted that the table does not, of course, contains all test procedures available (Table 18.1).

Table 18.1 Test procedure and example

Test	When is the test used?	Examples
Examination of a group		
One-sample t-test	Examination of a mean for a metric and normally distributed variable.	What is the average age of business founders?
Wilcoxon Test	Examination of a mean for a metric but non-normally distributed variable or an ordinal variable.	How much professional experience do founders have on average?
χ^2-test	Comparison of observed frequencies with theoretically expected frequencies for a nominal variable.	Does the proportion of female founders reflect the gender share in society?
Examination of two independent groups		
Independent samplest-test	Testing for a difference between groups for a metric and normally distributed variable.	Is there a difference in age between men and women when starting an enterprise?
Mann-Whitney test	Testing for a difference between groups for a metric but non-normally distributed variable or an ordinal variable.	Is there a difference in the professional experience between male and female founders?

(continued)

F. Kronthaler, *Statistics Applied With Excel*, https://doi.org/10.1007/978-3-662-64319-8_18

Table 18.1 (continued)

Test	When is the test used?	Examples
χ^2-test	Investigates whether two groups differ with regard to the frequency proportions of a nominal variable.	Do founders of service firms and those of industrial firms differ with respect to their motives for founding?
Examination with more than two independent groups		
Analysis of variance	Testing for a difference between more than two groups for a metric and normally distributed variable.	Do entrepreneurs differ in the workload by motives for starting the enterprise?
Examination of two dependent groups		
Dependent samples t-test	Examination of whether there are differences in a group before and after a measure for a metric and normally distributed variable.	Does an advertising measure have an impact on the image of my product (measured in a metric way)?
Wilcoxon test for dependent samples	Examine whether there are differences in a group before and after a measure, for a metric but non-normally distributed variable or an ordinal variable.	Does an advertising measure have an effect on the image of my product (measured ordinally)?
McNemar χ^2-test	Examine whether there are differences in a group before and after a measure for a nominal variable.	Does the advertising measure have an effect on whether or not people buy my product?
Testing for correlation between two variables or for independence		
Bravais-Pearson	Examine whether there is a relationship between two metric variables.	Is there a correlation between the expenditure on marketing and the expenditure on innovation?
Spearman	Examine whether there is a relationship between two ordinal variables.	Is there a relationship between healthy nutrition and education?
χ^2-Independence test	Investigates whether there is a relationship between two nominal variables.	Is there a relationship between the industry in which the enterprise is founded and the motive to found the business?

Regression Analysis: The Possibility to Predict What Will Happen

Regression analysis is a very popular tool that is applied to a lot of research questions. For example, we might ask whether expenditures on marketing have an effect on business growth and what happens if we increase or decrease expenditures. We could also try to find out which factors have an influence on the performance of football teams or which European football club is likely to win the Champions League: Barcelona, Bayern, Real Madrid or Liverpool? These examples show that the possible applications are manifold. In general, we use regression analysis to explain and predict. To illustrate this once more, here are a few possible questions:

- What factors determine the performance of a football team? Is the overall value of the football team or the presence of a superstar important?
- What impact will my marketing have on my sales figures? What happens if marketing expenditure increases?
- What impact does research and development have on a country's innovation rate? What happens when a country increases its spending on basic research?
- What effect does development aid have on the growth of developing countries? What happens when development aid are increased?
- What effect does violence on television have on young people? What happens when television shows become more violent?

There are many possible applications. Unfortunately, even with sophisticated statistical techniques, answering such questions is not always easy and, as we will see in a moment, requires thorough theoretical considerations in addition to statistical methods.

The Simple Linear Regression

19

19.1 Objectives of Regression Analysis

The objective of simple linear regression is to explain the effect of a variable X on a variable Y and what happens to variable Y when the variable X changes or is changed. Accordingly, Y is the dependent variable and X is the independent variable:

$$Y \longleftarrow X$$

dependent variable (metric)   *independent variable (metric)*

We can read the relationship as follows. The variable Y, the dependent variable, is influenced by the variable X, the independent variable. Both variables have metric scale level.

Freak Knowledge

In addition to the terms dependent and independent variable, other terms are used as well, depending on the field of study:

dependent variable	independent variable
endogenous variable	*exogenous variable*
response variable	*explanatory variable*
criterion variable	*predictor variable*
regressand	*regressor*

© Springer-Verlag GmbH Germany, part of Springer Nature 2023
F. Kronthaler, *Statistics Applied With Excel*,
https://doi.org/10.1007/978-3-662-64319-8_19

The division into dependent and independent variables is not a statistical, data-based one. It is theoretical in nature. That is, the division is based solemnly on the theoretical reasoning that X has an influence on Y. This theoretical foundation is central to regression analysis. If we do not have it, the result of the regression analysis is worthless. We cannot then say what happens to Y if we change X, because we do not know whether X really influences Y or whether the relationship is maybe the other way around.

Let's discuss this again with an example. The question is maybe as follows: Does the founder's job experience have an impact on turnover growth? To answer the question, we first use existing theories and literature to describe the relationship. We then possibly arrive at the research hypothesis that job experience has an impact on turnover growth. Perhaps founders with more professional experience know the market better and therefore can offer better products, etc. Hence, professional experience is the independent variable and turnover growth the dependent variable.

$$Y \longleftarrow X$$

turnover growth (metric)   *professional experience (metric)*

If we look more closely at the issue of causality, we notice that the direction here is very clear. Professional experience at the time of founding an enterprise can have an impact on turnover growth of the enterprise. The reverse is not possible, since turnover growth can only happen after the time of founding the enterprise. Turnover growth cannot influence the professional experience at the time point of founding the business. Unfortunately, the question of causality is not always that simple.

19.2   The Linear Regression Line and the Ordinary Least Squares Method

The objective of linear regression is, as the name indicates, to analyze a linear relationship between the dependent variable and the independent variable. The best way to explain the procedure is to use an example. Here, as discussed above, we define the growth rate of the enterprises as the dependent variable Y and the experience of the company founders as the independent variable X.

$$Y \longleftarrow X$$

growth rate (metric)   *experience (metric)*

We start with the scatterplot between the two variables. The scatterplot displays the relationship between the dependent and independent variables and gives information whether the relationship is linear in nature (Fig. 19.1).

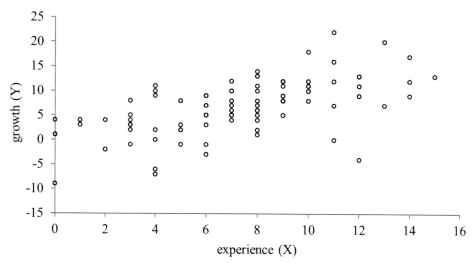

Fig. 19.1 Scatterplot for the variables growth rate and experience

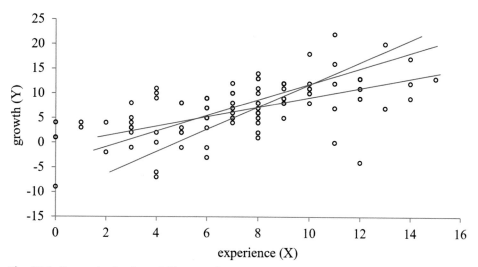

Fig. 19.2 Scatterplot for the variables growth rate and experience with optimal straight line?

The objective of simple linear regression is to put an optimal straight line through the point cloud. As with the Bravais-Pearson correlation coefficient, this assumes that the relationship is linear.

The question is how we can put a straight line in the point cloud so that it optimally describes the relationship. One way is to draw freehand the straight line to the best of our ability. In Fig. 19.2 three possibilities are drawn how the straight line could look like.

But which is the best line? It is difficult to decide visually. We need a better method. One of the best methods we know is the ordinary least squares method.

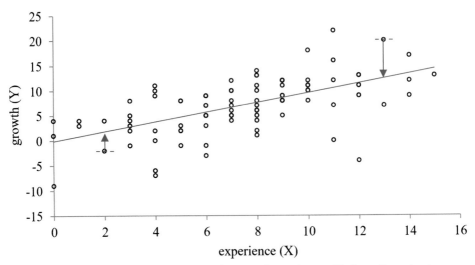

Fig. 19.3 Scatterplot for the variables growth rate and experience with the ordinary least squares line

Using the ordinary least squares method, we minimize the distance between our observations and the straight line we are looking for (Fig. 19.3). If this distance is minimal, we have found the optimal straight line.

Mathematically, a straight line can be written as follows:

$$Y = b_0 + b_1 X$$

where

b_0 is the intercept, i.e. the point of intersection of the line with the Y-axis, and

b_1 is the slope of the line.

The slope describes how much the variable Y changes when X increases by one (Fig. 19.4).

Since we are using the least-squares method to estimate the straight line, we write for the estimated line (we also refer to it as the regression line), as follows:

$$\hat{Y} = b_0 + b_1 X$$

where

$\hat{Y}$ (Y-hat) are the estimated Y-values,

b_0 is again the intercept, and

b_1 is the slope.

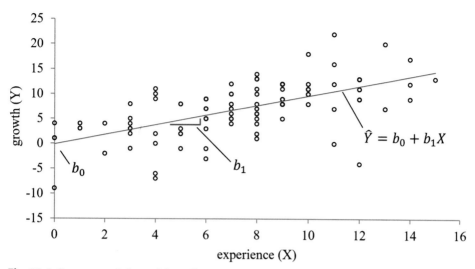

Fig. 19.4 Intercept and slope of the ordinary least squares line

Now we can turn to the minimization rule. The objective of the ordinary least squares method is to minimize the sum of the deviations between observed y_i values and estimated $\widehat{y}_i$ values. We can also write the deviation as follows:

$$e_i = y_i - \widehat{y}_i$$

where

e_i is the deviation between observed and estimated values.

If we sum up all the deviations, we arrive at the following formula:

$$\sum e_i = \sum (y_i - \hat{y}_i)$$

Additionally we can substitute $\widehat{y}_i$ by the straight line equation $\widehat{y}_i = b_0 + b_1 x_i$ and we receive

$$\sum e_i = \sum (y_i - (b_0 + b_1 x_i)).$$

Now one last step is necessary and we will understand why the method is called ordinary least squares method. The line we place in the point cloud using the method is the average straight line that best describes the point cloud. Since it is the average line, the sum of the deviations is always zero, we have positive and negative deviations. We therefore are not able to minimize deviations, we have to square them. The least squares rule is hence as follows:

$$\sum e_i^2 = \sum (y_i - (b_0 + b_1 x_i))^2 \rightarrow \min!$$

We find this minimum if we calculate the derivatives with respect to b_0 and b_1. When we do this, we receive the following formulas for the intercept and slope, which we can use to determine the regression line from our observations:

$$b_0 = \frac{\sum x_i^2 \sum y_i - \sum x_i \sum x_i y_i}{n \sum x_i^2 - \left(\sum x_i\right)^2}$$

$$b_1 = \frac{n \sum x_i y_i - \sum x_i \sum y_i}{n \sum x_i^2 - \left(\sum x_i\right)^2}$$

Alternatively, the slope and intercept can be calculated using the Bravais-Pearson correlation coefficient, the standard deviations, and the mean values. The formulas to do that in this way are as follows:

$$b_0 = \bar{y} - b_1 \times \bar{x}$$

$$b_1 = r \times \frac{s_y}{s_x}$$

with
$\bar{y}$ and $\bar{x}$ are the mean values of the dependent and the independent variable,
r is the Bravais-Pearson correlation coefficient between X and Y,
s_y and s_x are the standard deviations of X and Y.

It does not matter which method is used, the result should be the same.
    Let's now use the first eight companies in our data set as example to calculate our first regression. The best way to do this is to create a table and calculate the required values.

Enterprise	y_i	x_i	x_i^2	$x_i y_i$
1	5	7	49	35
2	8	8	64	64
3	18	10	100	180
4	10	7	49	70
5	7	6	36	42
6	12	10	100	120
7	16	11	121	176
8	2	4	16	8
Σ	78	63	535	695
		$\left(\sum x_i\right)^2 = 63^2$		

We insert the values according to the formulas and obtain the intercept and slope of the regression line:

$$b_0 = \frac{\sum x_i^2 \sum y_i - \sum x_i \sum x_i y_i}{n \sum x_i^2 - \left(\sum x_i\right)^2} = \frac{535 \times 78 - 63 \times 695}{8 \times 535 - 63^2} = -6.61$$

$$b_1 = \frac{n \sum x_i y_i - \sum x_i \sum y_i}{n \sum x_i^2 - \left(\sum x_i\right)^2} = \frac{8 \times 695 - 63 \times 78}{8 \times 535 - 63^2} = 2.08$$

With this we can formulate the regression line and calculate for certain X values estimated Y values.

$$\hat{Y} = b_0 + b_1 X = -6.61 + 2.08X$$

For example, we want to calculate the estimated growth rate for an enterprise whose founder has 11 years of professional experience. To do this, we simply insert 11 for our X we receive the estimated growth rate:

$$\hat{Y} = -6.61 + 2.08 \times 11 = 16.27$$

The growth rate estimated using the regression line is 16.27%. Remember, we had said that the slope is the change that occurs when we increase the X value by one. Let's check that once and calculate the estimated growth rate for 12 years.

$$\hat{Y} = -6.61 + 2.08 \times 12 = 18.35$$

The change is $18.35 - 16.27 = 2.08$, which is exactly the slope. As we see, we can determine the regression line relatively easily from our observed values and use it to explain the effect and make predictions. However, it is not quite that simple. First we have to ask how well our calculated regression line explains the relationship.

19.3   How Much do We Explain, the R^2?

The problem with the regression line is that we can compute an optimal line for any point cloud, no matter how much and how well the line fits. Figure 19.5 illustrates this. On the left side, we have a straight line that explains the relationship between X and Y fairly well, the points scatter closely around the straight line. On the right side is a straight line that explains the relationship poorly, the points are far away from the straight line. Since the points deviate a lot, we can only say with great uncertainty what will happen to Y if we change X.

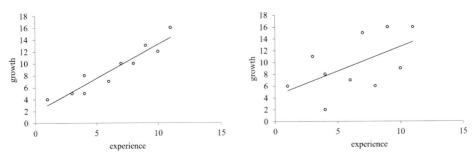

Fig. 19.5 Explanatory power of the regression line

The measure we use to determine the explanatory power of the regression line is the coefficient of determination R^2. R^2 indicates how much of the change of the dependent variable Y is explained by the independent variable X. Technically speaking, what percentage of the variance of Y is explained by X:

$$R^2 = 1 - \frac{\text{unexplained variance}}{\text{total variance}} = 1 - \frac{\sum (y_i - \hat{y}_i)^2}{\sum (y_i - \bar{y})^2}$$

Defined is the coefficient of determination between 0 and 1. 0 means that the regression line explains 0% of the change (variance) of Y. 1 means that it explains 100% of the change (variance) of Y. Accordingly, e.g., a R^2 of 0.3, means that 30 % of the variance of Y is explained by the regression line, the other 70% are, hence, not explained by the regression line.

Freak-Knowledge
The calculation of the R^2 is based on analyzing the variance. The deviation of the Y values of their mean can be divided into a part explained by the regression line and a part not explained by the regression line. If the explained part is high, the unexplained part is small, and vice versa.

To calculate the R^2, we must first calculate the mean $\bar{y}$ and the estimated $\hat{y}_i$ values. Then we determine the sum of the squared deviations of the y_i from $\bar{y}$ and from $\hat{y}_i$. Afterwards we insert all into the formula and yield the result. The result tells us what percentage of the variance of Y is explained by the regression line. Let's continue the example from above. Here we had estimated the following regression line:

$$\hat{Y} = b_0 + b_1 X = -6.61 + 2.08X$$

Let's put into the regression line our x_i values, so we get the estimated $\hat{y}_i$ values (see table) and, once we have also calculated the mean value, we can determine the required sums.

Enterprise	y_i	x_i	$\hat{y}_i$	$y_i - \hat{y}_i$	$(y_i - \hat{y}_i)^2$	$y_i - \bar{y}$	$(y_i - \bar{y})^2$
1	5	7	7.95	− 2.95	8.7025	− 4.75	22.5625
2	8	8	10.03	− 2.03	4.1209	− 1.75	3.0625
3	18	10	14.19	3.81	14.5161	8.25	68.0625
4	10	7	7.95	2.05	4.2025	0.25	0.0625
5	7	6	5.87	1.13	1.2769	− 2.75	7.5625
6	12	10	14.19	− 2.19	4.7961	2.25	5.0625
7	16	11	16.27	− 0.27	0.0729	6.25	39.0625
8	2	4	1.71	0.29	0.0841	− 7.75	60.0625
Σ	78				37.772		205.5
	$\bar{y} = 9.75$						

If we fill in the calculated sums into our formula, we get an R^2 in the amount of 0.82:

$$R^2 = 1 - \frac{\sum (y_i - \hat{y}_i)^2}{\sum (y_i - \bar{y})^2} = 1 - \frac{37.772}{205.5} = 0.82$$

Hence, with the help of the regression line we can explain 82 % of the variance of Y.

As a further example, let's calculate the R^2 for the left and right sides of Fig. 19.5. For this we first have to read the values from the figure. After we calculated both R^2 values we see that the value for the left side is 0.9 and for the right side 0.32, i.e. the regression line in the left side of the figure explains about 90% of the variance of Y and the regression line in the right side explains about 32 %.

In summary, the higher the R^2 the better the regression line fits to the data and the more accurately we can predict what will happen with Y when X changes. A high R^2 is thus an indication of the fit of the regression analysis. Nevertheless, in the social sciences we are often already content with a small R^2, e.g. in the amount of 0.2 or 0.3. The reason for this is that we often work with concepts that are difficult to measure and prone to error when measured. The data is therefore relatively often not really precise and we have something what is called noise in the data. The R^2 is then inevitably often relatively small. As we will see in Chap. 20, however, not only the R^2 is decisive for the quality of the regression line, but we can also test whether the regression line as a whole has explanatory power.

19.4   Calculating Simple Linear Regression with Excel

With the help of Excel, the calculation of the simple linear regression is very easy. We have two options. First, we can use Excel to draw the scatter plot and then display both the formula of the regression line and the coefficient of determination in the scatter plot. Second, we have the functions INTERCEPT, SLOPE, and RSQ at our disposal. Using our data set as an example, we can show this for all observed enterprises.

When we use the scatter plot, we first draw it and then click on the plot. A large "plus sign" appears, we click on the sign and a window opens offering the command "Trendline".

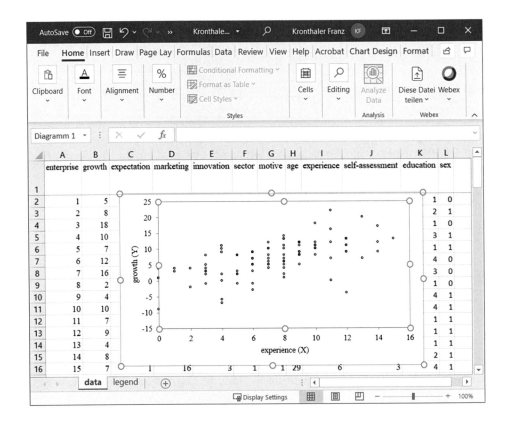

We click on the small arrow to the right of "Trendline" and we have the option to add a linear line.

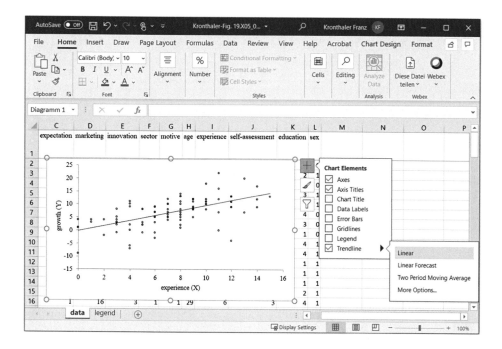

We select Linear and the least squares line is displayed in the scatterplot. Then we click a little below on "More Options" and scroll down until we can check "Display Equation on chart" and "Display R-squared value on chart". We close the box and both the formula of the regression line and the coefficient of determination are displayed in the diagram. We only have to note that in the regression equation the slope is displayed before the intercept.

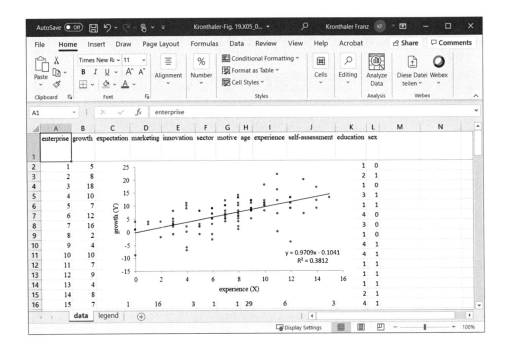

If we use the functions INTERCEPT, SLOPE, and RSQ, we call the corresponding functions via the Formulas tab and the Insert Function command, enter our Y and X data. Excel displays afterwards the intercept b_0, the slope b_1 and the coefficient of determination R^2.

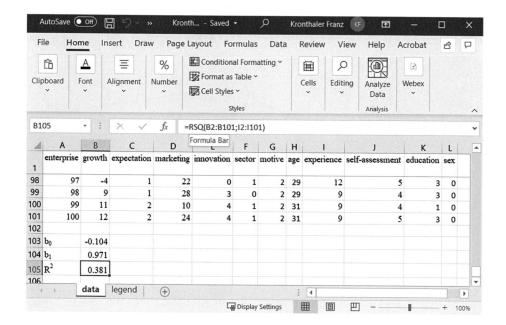

Accordingly, our regression line is $\widehat{Y} = -0.1041 + 0.9709X$ with an explanatory power of about 38%.

19.5   Is One Independent Variable Enough, Out-of-Sample Predictions, and Even More Warnings

To conclude the chapter, a few warnings and notes about the applicability of simple linear regression should be made.

One independent variable to explain the dependent variable is usually not enough. We can easily imagine that firm growth depends not only on professional experience of the founder, but that perhaps marketing or innovation also have an impact. Likewise, it could be that the sector is relevant and that it makes a difference whether the founder is a woman or a man. Also, the age of the founder or the motive for starting the business could be relevant. In other words, for each question, we usually have several independent variables that may influence the dependent variable. Accordingly, we must include all relevant variables in the regression model. If we do not do this, we may overestimate the influence of the included independent variable and our forecasts will be inaccurate.

Another problem with forecasts is that we should normally only make them in the range of our observations. If we make predictions outside our observation range, we speak of out-of-sample predictions. The main problem with out-of-sample predictions is that we must assume that the relationship we have found in our data also holds beyond the observed

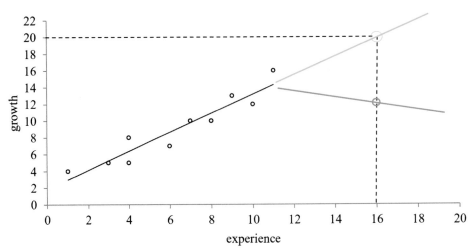

Fig. 19.6 Out-of-sample prediction

range of values. This may or may not be so. If it does not hold, then we are making an incorrect prediction. Let's compare the corresponding figure (Fig. 19.6).

In the figure, the observed range of values for the independent variable ranges from one to eleven. For this range of values, we have a close linear relationship between the variables experience and growth rate. If we use the relationship to forecast outside the observed range, we assume that the line can be continued. Thus, at 16 years, we arrive at the prediction of a growth rate of 20%. However, if the relationship changes, then the true value would not be 20%, but about 13%. Accordingly, for out-of-sample predictions, we need to have a good theoretical justification for why the observed trend may continue, further we need to be careful in interpreting the prediction.

One final comment. Linear regression computes a straight line. But this also means that there must be a linear relationship in the data. If it does not exist, linear regression is inadequate to represent the relationship.

In Fig. 19.7 there is no relationship between X and Y. Y does not change when X changes. However, we have an outlier in the upper right. The regression line takes this point into account and calculates a linearly ascending line accordingly.

In Fig. 19.8 two point clouds are shown. At the bottom left Y changes unsystematically with a change in X, and on the upper right this is also the case. So we do not have a linear relationship between X and Y, but we do have separate point clouds. The simple regression line tries to calculate a straight line between the point clouds and is accordingly not suitable to represent the data.

In Fig. 19.9 we see a non-linear relationship between X and Y. Linear regression attempts to calculate the relationship using a straight line, which is obviously not correct in this case.

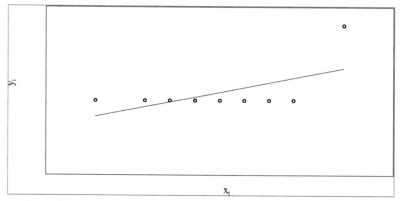

Fig. 19.7 No relationship between dependent and independent variable - #1

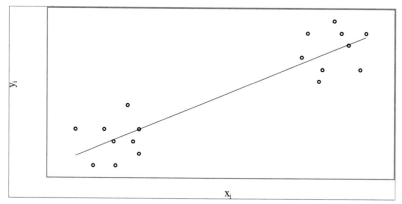

Fig. 19.8 No relationship between dependent and independent variable - #2

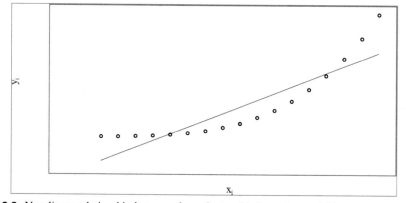

Fig. 19.9 Non-linear relationship between dependent and independent variable

Freak Knowledge
If there is a non-linear relationship between the variables, e.g. in the form of a u-shaped or an exponential relationship, there are mathematical transformation possibilities to transform such a relationship into a linear relationship. Linear regression can then be used again.

19.6 Checkpoints

- *Simple linear regression is used to estimate a linear relationship between two metric variables.*
- *Applying regression analysis requires a theory-driven approach. We need a sound theoretical foundation which shows that the independent variable has an impact on the dependent variable.*
- *The ordinary least squares method is used to estimate the regression line.*
- *The coefficient of determination R^2 is a measure of the strength of the relationship. It explains how much of the variance of the dependent variable is explained by the independent variable.*
- *Usually, one independent variable is not enough to explain a dependent variable, i.e., typically more than one independent variable is necessary.*
- *Out-of-sample predictions are based on the assumption that the discovered relationship holds beyond the observation range, and should be applied with caution.*

19.7 Applications

19.1 Why are theoretical considerations necessary before conducting a regression analysis?

19.2 For the first eight companies of our data set, data_growth.xlsx, draw and calculate by hand the scatterplot, the regression line, and the coefficient of determination for the variable growth and marketing. Use the variable growth as the dependent variable and the variable marketing as the independent variable. Interpret the result.

19.3 Using the result from application 2, predict the growth rate for marketing expenditures in the height of 20%. Is there a problem with the prediction?

19.4 For the first eight companies of our data set data_growth.xlsx, draw and calculate by hand the scatterplot, the regression line, and the coefficient of determination for the variable growth and innovation. Use growth as the dependent variable and innovation as the independent variable. Interpret the result.

19.5 Using the result from application 4, predict the growth rate for innovation expenditures in the height of 20%. Is there a problem with the prediction?

19.6 Using Excel, draw and calculate the scatterplot, regression line, and coefficient of determination for the following pairs of variables: growth and marketing, growth and innovation, and growth and age (data_growth.xlsx). Use the variable growth as the dependent variable. Interpret the results.

19.7 With task 6 in mind, describe why one independent variable is not sufficient to explain the growth rate.

Multiple Regression Analysis

20

20.1 Multiple Regression Analysis: More than One Independent Variable

In the previous chapter we discussed that usually one independent variable is not sufficient to describe the dependent variable. Usually, several factors influence the dependent variable. Hence, typically we need multiple regression analysis, also called multivariate regression analysis, to describe an issue. With multiple regression analysis we are able to simultaneously analyze the impact of several independent variables on a dependent variable.

$$Y \longleftarrow X_1, X_2, \dots X_J$$

dependent variable   *independent variables*

(metric)   *(metric and dummy-variables)*

The dependent variable is metric, the independent variables are also metric or so-called dummy variables. Dummy variable is the technical term for nominal variables with the expressions 0 and 1, e.g. man and woman, industrial enterprise and service enterprise, Catholic and non-Catholic.

> **Freak Knowledge**
> Nominal variables with more than two expressions, or ordinal variables, can be included in multiple regression analysis as well, but they have to be recoded into dummy variables.

© Springer-Verlag GmbH Germany, part of Springer Nature 2023
F. Kronthaler, *Statistics Applied With Excel*,
https://doi.org/10.1007/978-3-662-64319-8_20

The goal of multiple regression analysis is to describe the impact of a set of independent variables X_j on a dependent variable Y and to determine what happens to Y when a variable X_j changes or is changed. The difference with simple linear regression is we simply do not longer examining just one independent variable, but more than one independent variables simultaneously.

Although we analyze several independent variables simultaneously, we are still usually interested in only one of these variables. We include the other independent variables in the regression analysis because they also impact the dependent variable. To determine the true influence of the variable of interest, we must control for the influence of the other variables. Technically, therefore, we speak of control variables. We determine the influence of an independent variable on Y controlling for all other relevant variables.

Let's illustrate this again with an example. We are interested in the influence of professional experience on firm growth. The theory indicates that professional experience "'may'" (we have not read the theory, so only may) have a positive impact. At the same time, we read in the theory that not only professional experience could play a role for firm growth, but also marketing, research and development, the sector, etc. Therefore, to find out the real impact of professional experience, we need to control for these factors.

The procedure of estimating the regression function is identical to that of simple linear regression. Using the ordinary least squares method, we minimize the squared deviations of the regression function from the observed y_i-values:

$$\sum e_i^2 = \sum \left(y_i - \left(b_0 + b_1 x_{1i} + b_2 x_{2i} + \ldots + b_j x_{ji} \right) \right)^2 \rightarrow \min!$$

With the help of the minimization rule, we obtain the regression function with the estimated coefficients:

$$\hat{Y} = b_0 + b_1 X_1 + b_2 X_2 + \ldots + b_J X_J$$

where

b_0	is the intercept,
b_1 to b_J	are the estimated coefficients for the independent variables X_j,
J	is the number of independent variables,
$\widehat{Y}$	are the estimated values using the regression function.

Let's illustrate this with our example. Theory shows that in addition to professional experience, marketing and innovation, the sector, the age and gender of the founder all might impact enterprise growth. Hence, we have six independent variables to include in our regression model. Two of them, the sector and the gender, are dummy variables. We thus specify our model as follows:

$$\hat{Y} = b_0 + b_1 X_1 + b_2 X_2 + b_3 X_3 + b_4 X_4 + b_5 X_5 + b_6 X_6$$

respectively:

$$\hat{G} = b_0 + b_1 \text{Mark} + b_2 \text{Inno} + b_3 \text{Sec} + b_4 \text{Age} + b_5 \text{Exp} + b_6 \text{Sex}$$

where

$\hat{G}$ are the growth rates estimated using the regression model,
Mark   is the marketing effort,
Inno   is the effort for innovation,
Sec   is the sector in which the enterprise operates,
Age   is the age of the enterprise founders,
Exp   is the professional experience of the founders, and
Sex   is the gender of the founders.

Now, before we estimate this model using Excel (calculating it by hand is getting laborious), we need to briefly discuss the various measures used to determine the goodness of the fit of the regression model.

20.2 F-Test, t-Test and Adjusted-R^2

First, it should be clear that we are estimating the regression model from a sample. Hence, we receive the regression function for the sample. However, since we want to make a statement about the population, we need to test the regression model. We can test both the overall regression model and the regression coefficients for the independent variables. For the regression model, the null hypothesis and the alternative hypothesis are as follows:

H_0:   *The regression model does not explain the dependent variable.*
H_A: *The regression model explains the dependent variable.*

The significance level is usually $\alpha = 1\%$ resp. $\alpha = 0.01$.
If we reject the null hypothesis, we assume that the regression function explains the dependent variable in the population. The test statistic to be used is defined as follows:

$$F = \frac{R^2 / (J)}{(1 - R^2)/(N - J - 1)}$$

where

R^2 is the coefficient of determination,
J is the number of independent variables, and
N is the number of observations.

The test value is F-distributed with $df_1 = J$ and $df_2 = N - J - 1$ degrees of freedom.
The test, in short F-test, thus uses the explained variance relative to the unexplained variance. If the explained variance is large, F is large and we reject the null hypothesis. The F-test is the first thing we look at after calculating the regression model.

For each of the independent variables we have a separate null and alternative hypothesis:

H_0:   The independent variable X_j does not contribute to the explanation of the dependent variable.
H_A: The independent variable X_j contributes to the explanation of the dependent variable.

We specify the significance level in advance of the analysis with $\alpha = 10\%, 5\%, 1\%$ respectively $\alpha = 0.1, 0.05, 0.01$.

If we reject the null hypothesis, we assume that the independent variable X_j has explanatory power, using the following test statistic:

$$t = \frac{b_j}{s_{b_j}}$$

b_j is the estimated coefficient for the independent variable X_j,
s_{b_j} is the standard error for the estimated coefficient.

Freak Knowledge
Coefficients are estimated using one of many possible samples. Therefore, there are naturally many possible coefficients that vary around the true value in the population. The standard error represents this variation.

With the t-tests we check which of the variables has explanatory power. We can perform these tests before or after we have examined the coefficient of determination.

We already know about the coefficient of determination R^2. It compares the explained variance with the total variance and tells how much variance of Y is explained by the regression model.

$$R^2 = 1 - \frac{\text{unexplained variance}}{\text{total variance}} = 1 - \frac{\sum (y_i - \hat{y}_i)^2}{\sum (y_i - \bar{y})^2}$$

When analyzing the regression model, especially when comparing different regression models, we do not use the simple R^2, but a so-called "adjusted-R^2". It is to be interpreted in the same way as the simple R^2. The background is as follows: It is possible to increase R^2 by including more independent variables. For reasons we won't discuss here, adding another independent variable never leads to a reduction of the value of R^2. Thus, by adding independent variables, we can increase the R^2, and hence arbitrarily the explanatory power of the model. For this reason, we should us a measure that controls for the number of variables included. The measure is calculated as follows:

$$R^2_{\text{adj}} = R^2 - \frac{J \times (1 - R^2)}{N - J - 1}$$

As we see in the formula, the numerator becomes larger and the denominator becomes smaller the more variables we include in the regression model. Accordingly, the R^2_{adj} becomes smaller when more variables are included in the model, if the additional explanatory power of the variable included does not outweigh the effect. Once we have included all the relevant independent variables in the model, it does not matter whether we use the simple or the adjusted coefficient of determination for interpretation. The adjusted coefficient of determination is used primarily when we are uncertain about the number of independent variables to use. We can then use the R^2_{adj} to compare whether one or the other regression model should be used.

20.3   Calculating the Multiple Regression with Excel

We can now estimate and interpret the multiple regression model using Excel. Excel requires that all independent variables included in the regression function are in adjacent columns. Therefore, if necessary, we resave the data and delete all variables that we do not need (Fig. 20.1).

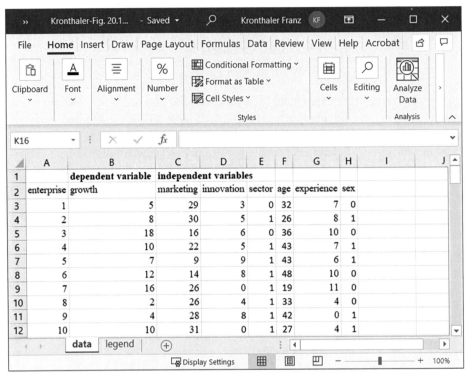

Fig. 20.1 Data structure when calculating multiple regression with Excel

Then, under the Data tab, we call up the command Regression using the Data Analysis tool and click OK.

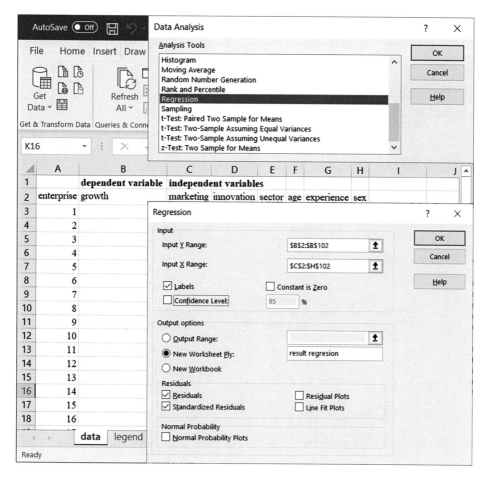

The data entry window appears. Here we enter the data. There are also several options available. We check the box for Labels when we enter the variables with a name. We have the result displayed in a new worksheet and check the boxes for residuals and standardized residuals. This way we have specified all relevant options. We then click OK and receive the result of our analysis (Fig. 20.2).

We can now turn to the interpretation of the results. In row 12 we find in column E the F-value and next to it in column F the probability of obtaining such a large F-value if the null hypothesis is correct. In our case, the probability is much smaller than 1%, i.e., we reject the null hypothesis. Accordingly, we assume the alternative hypothesis that the regression model explains the dependent variable. The R^2 in row 5 shows that the regression model explains about 48% of the variance of the dependent variable, which is already a relatively high value in social sciences. It is calculated from the values in rows 12, 13 and 14 of column C. Here we find the total variance of Y (line 14), the variance explained by the regression model (line 12), and the variance not explained by the regression model (line 13).

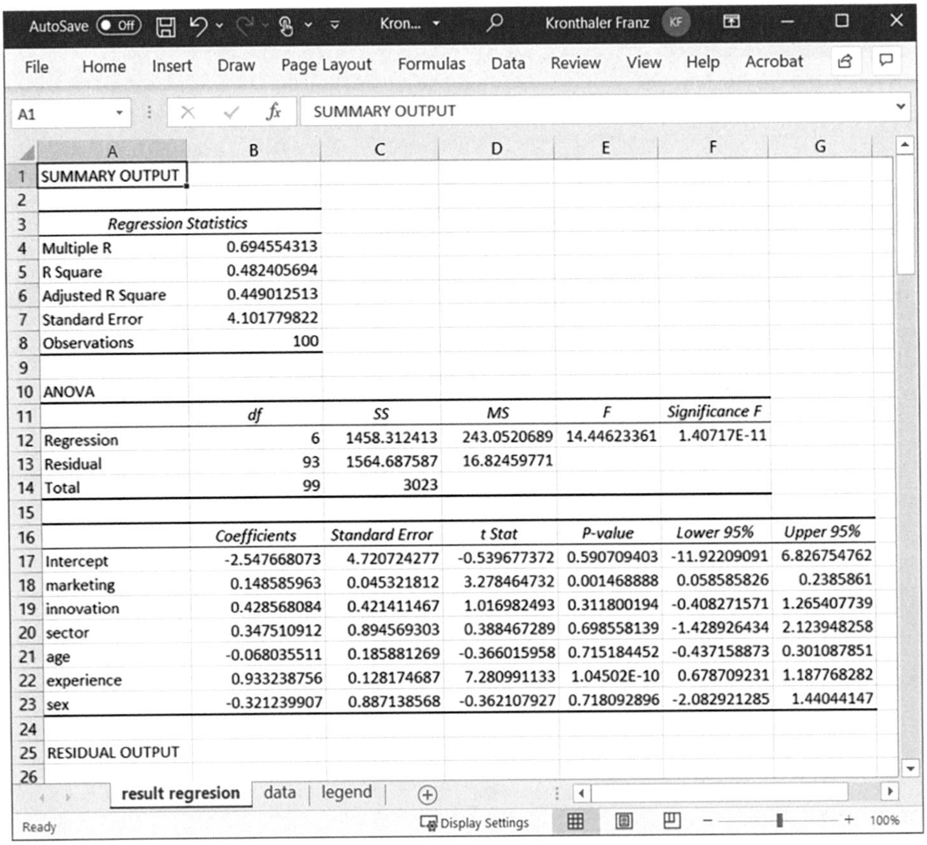

	A	B
1	SUMMARY OUTPUT	
2		
3	*Regression Statistics*	
4	Multiple R	0.694554313
5	R Square	0.482405694
6	Adjusted R Square	0.449012513
7	Standard Error	4.101779822
8	Observations	100
9		
10	ANOVA	

11		*df*	*SS*	*MS*	*F*	*Significance F*
12	Regression	6	1458.312413	243.0520689	14.44623361	1.40717E-11
13	Residual	93	1564.687587	16.82459771		
14	Total	99	3023			
15						

16		*Coefficients*	*Standard Error*	*t Stat*	*P-value*	*Lower 95%*	*Upper 95%*
17	Intercept	-2.547668073	4.720724277	-0.539677372	0.590709403	-11.92209091	6.826754762
18	marketing	0.148585963	0.045321812	3.278464732	0.001468888	0.058585826	0.2385861
19	innovation	0.428568084	0.421411467	1.016982493	0.311800194	-0.408271571	1.265407739
20	sector	0.347510912	0.894569303	0.388467289	0.698558139	-1.428926434	2.123948258
21	age	-0.068035511	0.185881269	-0.366015958	0.715184452	-0.437158873	0.301087851
22	experience	0.933238756	0.128174687	7.280991133	1.04502E-10	0.678709231	1.187768282
23	sex	-0.321239907	0.887138568	-0.362107927	0.718092896	-2.082921285	1.44044147
24							
25	RESIDUAL OUTPUT						

Fig. 20.2 Result of the calculation of the multiple regression with Excel

Now we can move to the individual independent variables. We find these in rows 18 through 23. We first focus on columns D and E. Here we find the t-value and the probability of finding such a large t-value if the null hypothesis is correct. For the variable marketing this probability is 0.15%, for innovation 31.18%, for the sector 69.86%, for age 71.52%, for experience 0.00% and for sex 71.81%. That is, we reject our null hypothesis for both the marketing variable and the experience variable. These variables help to explain the growth rate. For the other variables, we cannot reject the null hypothesis. In column B, we find the estimated coefficients that we use to formulate the regression function. Our regression function is:

$$\hat{G} = b_0 + b_1\text{Mark} + b_2\text{Inno} + b_3\text{Sec} + b_4\text{Age} + b_5\text{Exp} + b_6\text{Sex}$$

We can now add the estimated coefficients to these (rounded to two digits here):

$$\hat{G} = -2.55 + 0.15\text{Mark} + 0.43\text{Inno} + 0.35\text{Sec} - 0.07\text{Age} + 0.93\text{Exp} - 0.32\text{Sex}$$

Let's now take a closer look at the regression function. The first thing to note is that we formulate the regression function with all variables, regardless of whether they have explanatory power or not. This is always the case; we have, after all, estimated the regression function using these variables. b_0 is the value of the dependent variable when all independent variables are zero. In our case, the growth rate is -2.55% if we insert zero for all independent variables, e.g. zero percent of sales is spent on marketing, etc. The coefficients b_1 to b_6 indicate how much the growth rate changes if we increase the respective variable by one. This assumes that all other variables remain unchanged. The coefficient can only be applied to those variables that have explanatory power. In our case, these are the variables marketing and experience. For example, we can say what happens if we increase the variable marketing by one, the growth rate then increases by 0.15 percentage points. If the variable experience increases by one, the growth rate increases by 0.93 percentage points. In addition to the coefficients, in columns F and G we also find the confidence intervals within which the coefficients range with a 95% probability in the population.

20.4   When Is the Ordinary Least Squares Estimate BLUE?

We estimated the regression function using the ordinary least squares method. This method produces the best possible results when certain assumptions are met. We need to look at these assumptions and to discuss them. The following is a list of the assumptions:

1. The regression function is well specified and contains all the relevant independent variables. The relationship between the independent variables and the dependent variable is linear.
2. The deviations of the observed Y-values from the estimated $\widehat{Y}$-values have an expected value of zero.
3. The deviations are not correlated with the independent variables.
4. The variance of the deviations is constant.
5. Two or more independent variables are not correlated.
6. The deviations are uncorrelated with each other.
7. The deviations are normally distributed.

Under these conditions, the least squares estimator is BLUE, it is the best, linear, unbiased, and most efficient estimator we know.

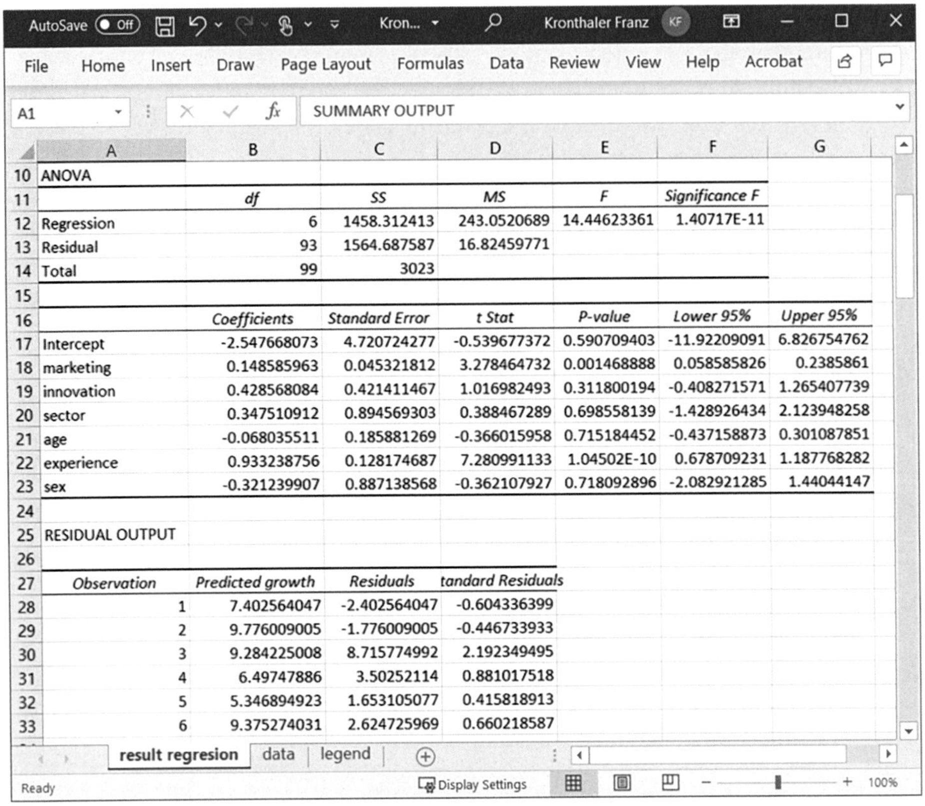

	A	B	C	D	E	F	G
10	ANOVA						
11		df	SS	MS	F	Significance F	
12	Regression	6	1458.312413	243.0520689	14.44623361	1.40717E-11	
13	Residual	93	1564.687587	16.82459771			
14	Total	99	3023				
15							
16		Coefficients	Standard Error	t Stat	P-value	Lower 95%	Upper 95%
17	Intercept	-2.547668073	4.720724277	-0.539677372	0.590709403	-11.92209091	6.826754762
18	marketing	0.148585963	0.045321812	3.278464732	0.001468888	0.058585826	0.2385861
19	innovation	0.428568084	0.421411467	1.016982493	0.311800194	-0.408271571	1.265407739
20	sector	0.347510912	0.894569303	0.388467289	0.698558139	-1.428926434	2.123948258
21	age	-0.068035511	0.185881269	-0.366015958	0.715184452	-0.437158873	0.301087851
22	experience	0.933238756	0.128174687	7.280991133	1.04502E-10	0.678709231	1.187768282
23	sex	-0.321239907	0.887138568	-0.362107927	0.718092896	-2.082921285	1.44044147
24							
25	RESIDUAL OUTPUT						
26							
27	Observation	Predicted growth	Residuals	tandard Residuals			
28	1	7.402564047	-2.402564047	-0.604336399			
29	2	9.776009005	-1.776009005	-0.446733933			
30	3	9.284225008	8.715774992	2.192349495			
31	4	6.49747886	3.50252114	0.881017518			
32	5	5.346894923	1.653105077	0.415818913			
33	6	9.375274031	2.624725969	0.660218587			

Fig. 20.3 Residuals output when computing multiple regression with Excel

> **Freak Knowledge**
> Unbiased means that the expected value of the coefficients is equal to the true values of the coefficients in the population. Efficient means that the coefficients are estimated with the smallest deviation possible.

We analyze most of these assumptions using the deviation of the observed Y-values from the estimated $\widehat{Y}$-values, also called residuals. We obtained these in our Excel result, see row 27 of Fig. 20.3.

In the following, we discuss the assumptions and how to test them using our example.

Re (1) The regression function must be well specified and linear in the relationship between the independent variables and the dependent variable. Well specified means that we have included all the relevant independent variables in the regression model. It requires that we have been theoretically sound in formulating the regression model. We can check whether there is a linear relationship between the dependent and independent variables by

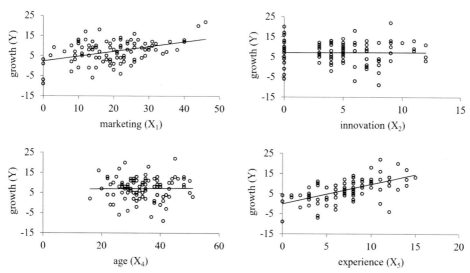

Fig. 20.4 Examining linearity between the dependent and the independent variables

using the scatterplots between the dependent and independent variables. To do this, we need to draw the scatterplots for all the metric independent variables and then visually examine whether there is a linear relationship between each independent variable and the dependent variable (Fig. 20.4).

The scatterplots drawn for the metric independent variables in our case show either a linear relationship (marketing and experience) or no relationship (age and innovation). A non-linear relationship, e.g. in the form of an arc cannot be identified. Therefore, we can conclude that the assumption of linearity is not violated. Scatterplots for our dummy variables sector and sex are not necessary, since the x-values can only take zero and one and thus no linear relationship between these variables and the dependent variable is possible.

Re (2) Condition two requires that the deviations of the observed Y-values from the estimated $\widehat{Y}$-values has the expected value of zero. This is a requirement which, if condition 1 is fulfilled, mainly concerns data quality. The observed Y-values should not be systematically measured too high or too low. If this is the case, then our intercept would be biased. If the Y-values are systematically measured too high, then our b_0 would be too high and vice versa.

Re (3) This assumptions requires that the deviations (I start talking about residuals) are not correlated with the independent variables. We can evaluate this assumption again with the scatterplot. We plot the residuals or the standardized residuals against the independent variables. If the condition is not violated, there should be no relationship between the variables. The residuals should scatter unsystematically. If we look at the scatterplots, this seems to be the case in our example. No trend or pattern can be observed (Fig. 20.5).

The assumption would be violated if, for example, we observe an ascending or a descending trend in the point cloud.

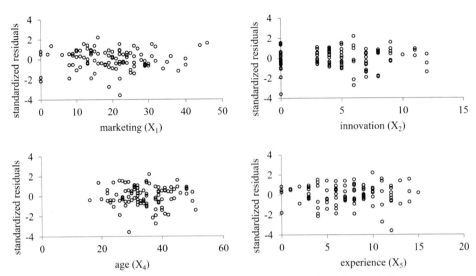

Fig. 20.5 Examining independency between residuals and the independent variables

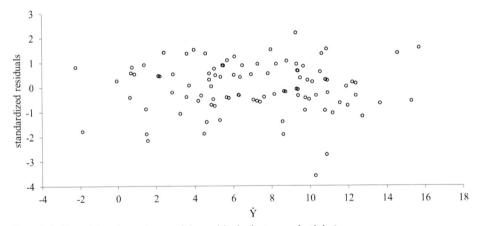

Fig. 20.6 Examining the variance of the residuals (heteroscedasticity)

Re (4) The fourth assumption requires that the variance of the residuals is constant. Technically, we speak of homoscedasticity or heteroscedasticity. Homoscedasticity means that over the range of values of the estimated $\widehat{Y}$-values, the variance is constant. Heteroscedasticity exists when the variance is not constant. We can evaluate the fourth assumption again with the help of our residuals. For this we draw the scatterplot for the residuals or the standardized residuals against the estimated $\widehat{Y}$-values. If the residuals scatter unsystematically in the same range, i.e., the deviations do not systematically become smaller or larger, then the assumption is satisfied. In our case, we see in the figure that the residuals scatter in approximately the same range (Fig. 20.6).

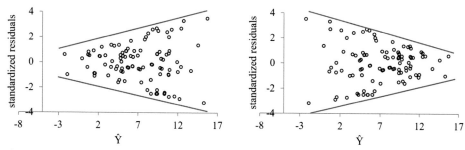

Fig. 20.7 Examples of heteroscedasticity

For example, if we find that the residuals funnels smaller or larger, then the assumption is violated (see Fig. 20.7) and the coefficients are no longer efficiently estimated. That is, the actual magnitude of the coefficients is more unknown than it needs to be.

Re (5) Assumption five requires that two or more independent variables are not correlated. Technically speaking, it requires that there is no multicollinearity between the independent variables. Using Excel, we have limited ability to test for multicollinearity. The easiest way is to calculate the correlation coefficients between the variables. If the correlation coefficient is higher than 0.8, we tend to have a problem with multicollinearity.

	Marketing	Innovation	Age	Experience
Marketing	1			
Innovation	−0.0312839	1		
Age	−0.0393392	0.954949346	1	
Experience	0.26496172	−0.266397792	−0.220079	1

In our case, the correlation coefficient between the variable innovation and the variable age is higher than 0.8, which means that we have a very high correlation between innovation and age. Thus, our ordinary least squares estimator can no longer properly assign the effect that the variables age and innovation have. The coefficients for the respective variables are potentially incorrect. The solution is to re-estimate the regression function and omit one of the highly correlated variables. In our case, we should remove either the variable age or the variable innovation. We need to consider which variable is more important.

Re (6) Assumption six requires that the residuals are not correlated with each other, that is, that successive residuals do not increase if the previous residual is large or decrease if the previous residual is large and vice versa. Technically speaking, this is the question of autocorrelation. We can analyze it again with the help of the scatterplot. We plot on the x-axis the observations one after the other, and on the y-axis the residuals. If the residuals

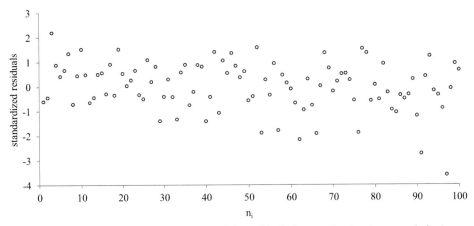

Fig. 20.8 Examination of the independence of the residuals from each other (autocorrelation)

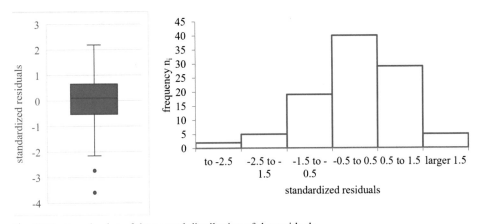

Fig. 20.9 Examination of the normal distribution of the residuals

scatter unsystematically and we do not recognize a pattern, the assumption is not violated. The following figure is a good example for this (Fig. 20.8).

Another hint: the problem of autocorrelation exists mainly when we analyze time series.

Re (7) The last assumption requires that the residuals are normally distributed. This assumption is especially important when we have a small sample. If we have a small sample and the assumption is violated, then our F-test and our t-tests are no longer valid and we do not know if we can actually reject the null hypotheses. We can evaluate the assumption visually with the help of the boxplot and the histogram (Fig. 20.9).

If the median in the boxplot deviates a lot from the center of the box, then the distribution is either left skewed or right skewed and there is a problem with the assumption. The histogram should look approximately bell-shaped. In our case, the distribution is slightly left skewed, but does not deviate much from a normal distribution. Based on the visual impression, the assumption does not seem to be seriously violated.

If any of these assumptions are seriously violated, our regression results are no longer trustworthy. The ordinary least squares method is then no longer BLUE, and we know of other procedures that give better results. The discussion of these procedures goes far beyond a basic statistics book. At this point I would like to refer to the more advanced statistical literature.

20.5 Checkpoints

- *With the help of multiple regression, we analyze the influence of several independent variables on a dependent variable.*
- *The dependent variable is metric, and the independent variables are metric or dummy variables.*
- *The F-test tests whether the whole regression model explains the dependent variable.*
- *The t-tests test whether which independent variable has explanatory power.*
- *The least squares method of estimating the regression function is BLUE if certain assumptions are met.*
- *If one or more of the assumptions are violated, then there are estimation procedures that give better results than the ordinary least squares method.*

20.6 Applications

20.1 Calculate the regression function from this chapter without the independent variable age.

20.2 Test the assumptions for the multiple regression of application 1.

20.3 By omitting the variable age, have we solved the problem of multicollinearity?

Part VI

What Happens Next?

How to Present Results

21.1 Contents of an Empirical Paper

After analyzing data, we usually have to write down the results, in the form of a student research paper, a bachelor thesis, a master thesis, or a journal article. Thereby the following points should always be discussed:

1. the problem,
2. the research question,
3. the literature,
4. the data and the method,
5. the empirical results,
6. the summary and the conclusions.

In the discussion of the problem, the reason and the motivation why dealing with the topic is addressed. If possible, there is a hook in the form of a current political discussion, etc. The research question is directly derived from the problem. In this way, the reader of the report understands why this research question was chosen. Subsequently, the state of knowledge is presented in a literature section and, if necessary, a research hypothesis is formulated with reference to the research question and the literature. Afterwards, the method and the data have to be discussed, where they come from, how reliable and valid they are and why the specific method was chosen. The next step involves the presentation of the empirical results. Finally, the results are to be summarized, related to the literature and it has to be discussed what conclusions can be drawn. To illustrate this, we will use two examples for different questions in the following. It should be remembered here that these are not full research reports, but only 'fictional' very brief reports. No real literature

© Springer-Verlag GmbH Germany, part of Springer Nature 2023
F. Kronthaler, *Statistics Applied With Excel*,
https://doi.org/10.1007/978-3-662-64319-8_21

review was conducted and the results are based on simulated data. This means that both the results and the reports have nothing, absolutely nothing, to do with reality.

21.2   Example I for a Report: Is There a Difference in Founding Age Between Male and Female Founders (Fictitious)

In the current socio-political discussion, it is problematized whether women and men differ in terms of age when founding an enterprise (see e.g. daily newspaper of 12.01.2015). Against this background, the aim of the present analysis is to determine whether this actually corresponds to the facts. Specifically, the question is analyzed whether male and female founders differ in their age at the time of founding an enterprise.

On the one hand, there are indications in the literature that women are older than men at the time of founding an enterprise. It is argued that women take longer to build up the necessary work experience to start a business. They interrupt their professional career more often, e.g. to take a maternity leave (see Author X 2014). Contrary to this, it is argued that women are more decisive, more willing to take entrepreneurial risks and therefore tend to start a business earlier than men (cf. Author Y 2012). [. . .]

[...] Therefore, since theoretically there are different opinions, we want empirically test the following hypothesis at the 10% significance level: Female and male founders differ in their age at the time of starting an enterprise.

To test the hypothesis, we draw on an existing representative data set on start-ups. The data set was collected in June 2013. Simple random sampling was used to draw a sample of 400 enterprises founded in 2007. The response rate was 25%, so the data set contains 100 enterprises. The data set informs with regard to the research question about the gender and age of the founders among other firm characteristics.

The test variable age is metric and normally distributed, accordingly the independent samples t-test can be applied. The following table contains the test result with equal variances.

Test variable: Age (in years) by gender			
Male founders	$\bar{x} = 34.77$	$s = 7.55$	$n = 65$
Female founders	$\bar{x} = 33.29$	$s = 7.76$	$n = 35$
$t = 0.93$	$df = 98$	$t_{crit} = \pm 1.66$	$P(T \leq t) = 0.36$
Effect size $r = 0.09$			

The result clearly shows that based on our sample finding, we cannot reject the null hypothesis that there is no difference in founding age between male and female founders. The probability of no difference in the population is relatively high. Moreover, the effect size indicates a small influence at best and the difference found in the sample is also relatively small with about 1.5 years. For the population, we therefore assume that male and female founders are of the same age at the time of founding and that there is no difference between the two groups.

21.3   Example II for a Report: Professional Experience and Enterprise Performance (Fictitious)

In the current economic policy discussion, there is increasing pressure to restrict the support of start-ups to those founders who have sufficient professional experience (see e.g. daily newspaper of 22.11.2013). The proponents of this measure hope that only promising companies will be subsidized and that state support will become more efficient and effective.

Against this background, the aim of this paper is to analyze the relationship between professional experience of founders and the growth of newly founded enterprises. Specifically, the following question is explored: What is the impact of professional experience on firm growth in the first five years of the firm's existence?

The theory and literature on professional experience and firm growth assume that professional experience has a positive influence on the growth of young firms. A founder with professional experience has more knowledge about the market and thus is better able to identify market opportunities (Author X 2011). Furthermore, the literature indicates that founders with professional experience obtain loans from banks and venture capital more easily. Since they have more capital available for investment, they are able to grow faster (Author Y 2009). [...]

[...] In summary, the following hypothesis can be stated: The amount of professional experience at the time of foundation has a positive impact on the turnover growth of newly founded enterprises.

To test this hypothesis, primary data was surveyed in June 2013. Using simple random sampling, a sample of 400 firms was drawn from firms founded in 2007. These companies were interviewed about various characteristics required for the study. Out of the 400 companies, 100 participated in the survey, representing a response rate of 25%. Various tests conducted indicate that the sample is representative.

The dependent variable employed is the average enterprise growth from 2007 to 2012. The independent variable of interest is the professional experience of the enterprise founders at the time point of founding the business. In addition, various control variables are implemented as it can be assumed that enterprise growth is correlated with other variables. Control variables are the expenditures for marketing and for innovation, the age of the founders, the sector and the gender of the founders. Overall, the following regression function is estimated using the ordinary least squares method:

$$\hat{G} = b_0 + b_1 \text{Mark} + b_2 \text{Inno} + b_3 \text{Sec} + b_4 \text{Age} + b_5 \text{Exp} + b_6 \text{Sex}$$

The tests on the assumptions of the ordinary least squares method show that there tends to be a problem with multicollinearity. The variables age and innovation are highly correlated with a correlation coefficient of 0.95. Therefore, to take the problem into account, the variable age is removed and the following reduced model is estimated:

$$\hat{G} = b_0 + b_1\text{Mark} + b_2\text{Inno} + b_3\text{Sec} + b_4\text{Exp} + b_5\text{Sex}$$

The estimation results show that we can reject the null hypothesis of the model. The model explains 48% of the variance of the dependent variable (see the following table).

	Ordinary least squares estimation; dependent variable: turnover growth
Intercept	-4.18***
Marketing	0.15***
Innovation	0.28**
Sector	0.33
Experience	0.93***
Sex	-0.28
Number of observations	100
R^2	0.48
F-test	17.47***

*, **, *** indicates statistical significance at the 10, 5, and 1% levels

Looking at the independent variable of interest, professional experience, we see that it is statistically significant at the 1% level.

The results of the study are consistent with the theoretical findings and existing literature. The founders' professional experience at the time point of the enterprise foundation has a positive effect on the firm growth of the newly formed enterprises. A caveat to note is that the research did not determine what specifically the higher growth was due to. It remains open whether the higher growth can be attributed to better market knowledge or whether it is due to better financing opportunities in the initial stage of the start-up. Before any final policy conclusions can be drawn, this issue needs to be addressed in further studies.

21.4 Applications

Select five journal articles in your field that use data and test them for the content described above.

Advanced Statistical Methods

22

With multiple regression analysis we discussed one of the most important multivariate analysis methods. However, it is not the only one. In addition, there are other multivariate analysis methods that are used for specific questions. We will give a brief outline of the most important ones here. Then we will turn in the next section to further statistical literature, which can be used to depend the statistical knowledge for example with regard to regression analysis and other multivariate analysis methods.

Logistic Regression

Logistic regression is very similar to multiple regression. The influence of several independent variables on a dependent variable is examined. The difference is that in the case of logistic regression, the dependent variable has only the expressions 0 and 1, an event does not occur or an event occurs. The applications are manifold, e.g. it can be analyzed, which factors were decisive for surviving or not surviving the Titanic disaster. Was the gender important, the age or the class in which one travelled? The question may also be what factors cause a loan to fail or not to fail. Another example is to study the purchase decision of individuals. What factors lead people to buy or not buy a particular product.

Discriminant Analysis

Discriminant analysis is used to analyze group differences. The results are often used to make a prediction about to which group new people or objects are likely belong to. For example, banks try to divide borrowers into bad risks (potential loan default) and good risks (loan not likely to default) in advance of granting a loan. For this purpose, the differences between good risks and bad risks are determined and new customers are classified based on these results.

© Springer-Verlag GmbH Germany, part of Springer Nature 2023
F. Kronthaler, *Statistics Applied With Excel*,
https://doi.org/10.1007/978-3-662-64319-8_22

Factor Analysis

We can use factor analysis to analyze the relationships between a larger number of metric variables. We are able to discover dimensions behind the correlation of variables, and can use factor analysis to reduce the number of variables. An example of this would be trying to measure the quality of restaurants. We can measure the quality of restaurants using different variables, such as taste, temperature of the dish, freshness, waiting time, cleanliness, friendliness of the employees, etc. Then, using factor analysis, we could analyze whether the number of variables can be reduced and whether there may be a common dimension behind the different variables such as quality of food or quality of service.

Cluster Analysis

Cluster analysis aims to group objects or people based on certain criteria. It attempts to group objects or people together in such a way that they are as similar as possible and different from the other groups. For example, one could identify consumer types in terms of price sensitivity, quality sensitivity, sustainability sensitivity, etc., or one could attempt to classify regions based on certain criteria such as gross domestic product, unemployment, infrastructure, industrialization, etc.

Conjoint-Analysis

Conjoint-Analysis is a tool that can be used to analyze which characteristics and which combinations of characteristics determine the benefit of a product or service. The user can use the procedure to analyze, for example, which attributes of a product, such as color, price, quality, size, etc., lead to a selection decision being made and which preferences prevail among consumers. Producers can use this knowledge to design products that meet consumer preferences.

Interesting and Advanced Statistical Textbooks

23

Finally, I would like to give some hints on interesting and advanced statistical textbooks. This section should be a little help for those who want to deal with statistics in more depth. But of course it cannot replace the own search for suitable literature. Beside the books mentioned, there are a number of other good textbooks. The selection contains German and English textbooks. I thought of leaving out the German books in the English version of this book, however, I decided not to do so. The books displayed here are in my opinion very good books and may some of you are interested in German books as well.

Döring, N.; Bortz, J.; unter Mitarbeit von S. Pöschl (2016), Forschungsmethoden und Evaluation in den Sozial- und Humanwissenschaften, 5. Aufl., Springer: Heidelberg
"Forschungsmethoden und Evaluation" is the German classic for questions on the research process, quantitative and qualitative data collection, hypothesis generation and hypothesis testing.

Bortz, J.; Schuster, Ch. (2010), Statistik für Human- und Sozialwissenschaftler, 7. Aufl., Springer: Berlin Heidelberg
"Statistik für Human- und Sozialwissenschaftler" is an outstanding presentation of descriptive statistics, inferential statistics and multivariate analysis methods. Like the book "Forschungsmethoden und Evaluation" it is a German classic of quantitative data analysis.

Backhaus, K.; Erichson, B.; Plinke, W.; Weiber, R. (2018), Multivariate Analysemethoden, 15. Aufl. Springer: Berlin Heidelberg
The book "Multivariate Analysemethoden" provides an in-depth introduction to multivariate data analysis methods. The reader will find a comprehensible presentation of regression analysis, time series analysis, analysis of variance, factor analysis, cluster analysis, etc. All methods are explained using data sets and the help of the statistical software SPSS.

© Springer-Verlag GmbH Germany, part of Springer Nature 2023
F. Kronthaler, *Statistics Applied With Excel*,
https://doi.org/10.1007/978-3-662-64319-8_23

Greene, W.H: (2018), Econometric Analysis, 8th Edition, Pearson: New York
"Econometric Analysis" is the standard work on the economic application of regression analysis. In over 1000 pages, the reader will find almost all the problems and solutions that may arise when doing regression analysis in the context of social sciences. The reader needs a deeper understanding of the mathematics.

Hair J.F.; Black W.C.; Babin, B.J.; Aderson, R.E. (2018), Multivariate Data Analysis, 8th Edition, Pearson: Harlow, UK
"Multivariate Data Analysis" is a simple and understandable textbook on multivariate data analysis procedures. The authors succeed in explaining the concepts and problems of the methods even without a lot of mathematics and apply them to examples. After reading the book, the reader is able to apply factor analysis, regression analysis, cluster analysis and other multivariate methods in depth. The chapter on data preparation is particularly noteworthy.

Gujarati, D. (2015), Econometrics by Example, 2nd Edition, Palgrave Macmillan: Basingstoke
For those who want to delve into regression analysis, the book "Econometrics by Example" is recommended. The book shows how to answer the central questions of regression analysis in a simple way using examples. It shows how to deal with problems without using a lot of mathematics. It is a book that non-technical people have been waiting for a long time.

Noelle-Neumann, E.; Petersen, T. (2005), Alle nicht jeder: Einführung in die Metho-den der Demoskopie, 4. Aufl., Springer: Berlin Heidelberg
Finally, a German classic of demoscopic research should be introduced: "Alle nicht jeder" explains in a simple and understandable way the problems of opinion research, interviewing, questionnaire construction, sampling, fieldwork and data analysis. The book is an excellent bedtime and holiday read for those who have always wanted to understand how pollsters find out who will be elected next Sunday.

Another Data Set to Practice

24

A key feature of the book is the application of the methods discussed to a data set. Hence, easily another data set could be used. To illustrate this, another data set is introduced in this section, which can be used to practice. Lecturers have the option to use this data set in class instead of the data set used in the book if it is more appropriate to their subject area. The complete simulated data set is available at www.statistik-kronthaler.ch. More data sets will be made available on this platform over time, which can be used in different fields of studies.

Let's imagine we are an intern in the company WebDESIGN. The company creates websites and internet solutions for other companies. WebDESIGN did a customer satisfaction survey in which 120 customers of WebDESIGN were randomly interviewed. As an intern we are given the task to analyze the data and find out how satisfied the customers are and what influences customer satisfaction. The following tables show the data for the first twelve customers as well as the legend (Tables 24.1 and 24.2).

From WebDESIGN's point of view, the data set can be used to analyze various questions. The following questions are intended to facilitate an introduction to the analysis of the data set. Beyond that, other questions are also imaginable:

- What is the number of observations? How many metric, ordinal and nominal variables does the data set contain?
- Using Excel, calculate the average values for all variables in the data set and interpret the results.
- For the data set, calculate the measures of variation for all variables and interpret the results.
- Draw the corresponding boxplots for the metric and ordinal variables and interpret the results.

© Springer-Verlag GmbH Germany, part of Springer Nature 2023
F. Kronthaler, *Statistics Applied With Excel*,
https://doi.org/10.1007/978-3-662-64319-8_24

Table 24.1 Data set WebDESIGN

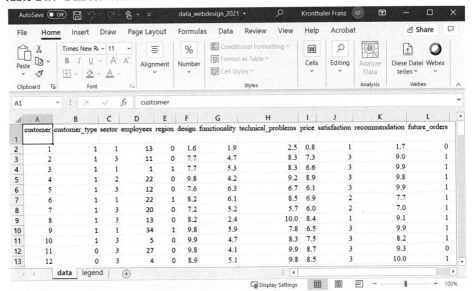

- Create the histograms for the variables design and functionality. Which of the two variables is more symmetric?
- Draw the scatterplots and calculate the Bravais–Pearson correlation coefficient for the variables recommendation and employees, recommendation and design, recommendation and functionality, recommendation and technical_problems, and recommendation and price. Interpret the results.
- Calculate Spearman's correlation coefficient for the variables recommendation and satisfaction and interpret the result.
- Calculate the phi coefficient for the variables customer_type and future_orders. Interpret the result.
- Calculate the contingency coefficient for the variables sector and future_orders. Interpret the result.
- What is the average size of our clients' companies and what is the range of the 90%-, 95%- and 99% confidence intervals?
- WebDESIGN is interested in finding out whether new customers or regular customers would rather recommend the company. Test this at the 10% significance level, considering all relevant steps.
- In addition, WebDESIGN wants to know if there is a difference between Swiss customers and German customers regarding recommendation. Test this at the significance level of 5%, considering all relevant steps.
- Test the already calculated correlation coefficients at the significance level of 5% considering all relevant steps.

Table 24.2 Legend for the data set WebDESIGN

	A	B	C	D	E
1	variable name	variable description	values	missing values	scale
2	customer	index questionnaire			
3	customer_type	type of customer	0=new customer 1=regular customer	n.d.	nominal
4	sector	sector in which the customer operates	1=industry 2=trade 3=service sector	n.d.	nominal
5	employees	number of employees of the customer	in full-time equivalents	n.d.	metric
6	region	region in which the company is located	0=Switzerland 1=Germany	n.d.	nominal
7	design	satisfaction with the design of the website		n.d.	metric
8	functionality	satisfaction with the functionality of the website	rating on a line from 0	n.d.	metric
9	technical_problems	satisfaction with the fixing of technical problems	(very unsatisfied) to 10	n.d.	metric
10	price	satisfaction with the price-performance ratio	(very satisfied)	n.d.	metric
11	satisfaction	overall satisfaction with webdesign	1=very unsatisfied 2=unsatisfied 3=satisfied 4=very satisfied	n.d.	ordinal
12	recommendation	it is likely that you recommend webdesign to other companies	rating on a line from 0 (very unlikely) to 10 (very likely)	n.d.	metric
13	future_orders	do you consider webdesign for future orders	0=no 1=yes	n.d.	nominal
14	Source: Own survey 2021.				
15	© Kronthaler, 2021				

- Calculate the regression function for the dependent variable recommendation with the independent variables customer_type, employees, region, design, functionality, technical_problems and price. Test whether the assumptions of the ordinary least squares method are violated. Perform the estimation again if necessary.

Solutions to the Applications

A

Chapter 1

1.1 First, with statistics knowledge is generated and we can generate knowledge ourselves. Second, statistics enables us to make informed decisions. Third, with knowledge of statistics, we can better evaluate studies and statements based on data and recognize attempts of manipulation.

1.2 In a data set, the following three main pieces of information are contained: for what objects or people do we have information; what information do we have; the information itself.

1.3 Nominal, ordinal and metric: a nominal scale allows us to distinguish people or objects; an ordinal scale gives information about the rank order in addition to the distinction; a metric scale allows to distinguish people or objects, rank them, and provides information about the difference between them.

1.4 The variable sector is nominal, the variable self-assessment is ordinal and the variable turnover is metric.

1.5 The variable education with the characteristics secondary school, A-level, bachelor, master is ordinal, a ranking is possible.

1.6 One possibility is to measure the number of years spent in an educational institution. An objection could be that the number of years spent in educational institutions has little to do with the level of education. The objection is a good one, but it can be raised in the same way if we measure educational attainment with school-leaving level. With a variable we try to describe a fact as close as possible.

1.7 The solution is the created data set in the task.

1.8 The scale of a variable determines, in addition to the question, which statistical procedure can be used.

1.9 A legend is necessary because the data are usually coded. With the help of the legend we understand the meaning of the data, even at a later point in time.

1.10 The data set contains 100 observations. It contains three nominal, two ordinal and five metric variables (the running index was not counted).

1.11 The trustworthiness of the source is decided by the seriousness of the source and our level of knowledge about it. data_growth.xlsx is not a real data set, but a simulated one for teaching purposes.

1.12 A solution can be found in Chap. 8.

Chapter 3

3.1 The mode is used with nominal, ordinal and metric data. Calculating the median is meaningful with ordinal and metric data. Calculating the arithmetic mean requires metric data. The geometric mean requires metric ratio scaled data.

3.2 $\bar{x}_{innovation} = 6\%$; $\bar{x}_{marketing} = 20\%$. The mean value is higher for the variable marketing, i.e. the enterprises spend on average a higher proportion of turnover on marketing.

3.3 $me_{self\text{-}assessment} = 4$; $me_{education} = 1.5$. The median for the variable self-assessment has a value of 4. 50% of the companies have a higher value, 50% a lower value. The median value for the variable education is 1.5. Fifty percent of the companies have a higher value, 50% a lower value.

3.4 $mo_{sex} = 0$ & 1; $mo_{expectation} = 2$. The most frequent values for the variable sex are 0 and 1. We have an equal number of male and female founders in the sample. The most frequent value for the variable expectation is 2. Most of the enterprises do not expect any change in the future development.

3.5 No, the variable is of nominal scale.

3.6 Results:

	A	B	C	D	E	F	G	H	I	J	K	L
1	enterprise	growth	expectation	marketing	innovation	sector	motive	age	experience	self-assessment	education	sex
95	94	8	1	29	0	1	1	27	9	4	1	0
96	95	3	1	19	9	1	3	43	3	2	4	0
97	96	-2	1	20	4	0	2	32	2	1	1	1
98	97	-4	1	22	0	1	2	29	12	5	3	0
99	98	9	1	28	3	0	2	29	9	4	3	0
100	99	11	2	10	4	1	2	31	9	4	1	0
101	100	12	2	24	4	1	2	31	9	5	3	0
102	mode	8.00	1.00	22.00	0.00	1.00	2.00	32.00	8.00	4.00	1.00	0.00
103	median	8.00	1.00	20.00	5.00			34.00	8.00	4.00	2.00	
104	mean	7.10		19.81	4.65			34.25	7.42			
105		metric	ordinal	metric	metric	nominal	nominal	metric	metric	ordinal	ordinal	nominal

3.7 The average growth factor is 1.18, and the average growth rate is 18%.

3.8 The average growth rate is around 10%. Based on a rounded growth factor of 1.10, the predicted revenue for 2016 is Fr. 283'232.

3.9 According to World Bank figures, India grew at an average rate of 1.096% and China at 0.516% during this period. If a prediction is made on this basis, then India (1'459'865'193) will have in 2025 for the first time a higher population than China (1'443'820'867).

3.10 The arithmetic mean is sensitive to outliers, all values are included in the calculation. The median and mode do not take extreme values into account, so they are not sensitive to outliers.

Chapter 4

4.1 The range and interquartile range can be used with ordinal and metric data. The standard deviation and variance require metric data.

4.2 Nominal data do not require a measure of variation because the data do not scatter around a value.

4.3 $ra_{age} = 29.0$; $s_{age} = 9.62$; $var_{age} = 92.57$; $iqr_{age} = 14.0$; $cv_{age} = 27.49$.

4.4 $ra_{marketing} = 21.0$; $s_{marketing} = 7.67$; $var_{marketing} = 58.86$; $iqr_{marketing} = 12.5$; $cv_{marketing} = 35.68$.

4.5 $ra_{innovation} = 9.0$; $s_{innovation} = 2.83$; $var_{innovation} = 8.00$; $iqr_{innovation} = 3.5$; $cv_{innovation} = 56.67$.

4.6 The coefficient of variation is the required measure of variation. The variation of each firm from the mean is larger for the variable innovation.

4.7 $ra_{Novartis} = 5.51$; $s_{Novartis} = 2.08$; $var_{Novartis} = 4.34$; $cv_{Novartis} = 2.85$; $ra_{UBS} = 1.27$; $s_{UBS} = 0.50$; $var_{UBS} = 0.25$; $cv_{UBS} = 6.25$. According to the coefficient of variation, the UBS stock was more volatile during this period.

4.8-9   Boxplot growth rate, the middle 50% of observations range from 4% (first quartile) to 11% (third quartile).

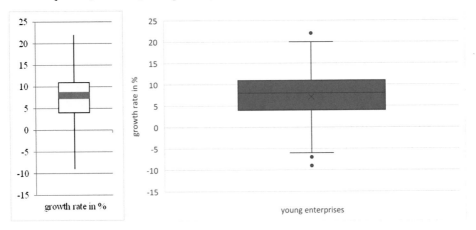

4.10-11   Boxplot growth rate by motive for starting a business (created with Excel).

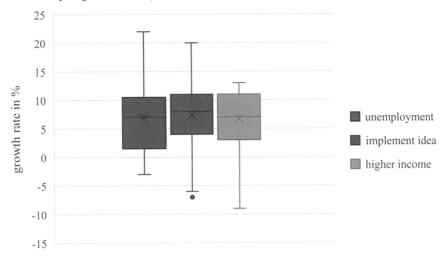

4.12   Results: The figure also shows the relationship between scale, averages and measures of variation.

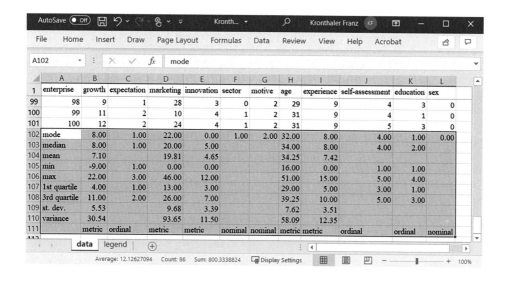

	A	B	C	D	E	F	G	H	I	J	K	L
1	enterprise	growth	expectation	marketing	innovation	sector	motive	age	experience	self-assessment	education	sex
99	98	9	1	28	3	0	2	29	9	4	3	0
100	99	11	2	10	4	1	2	31	9	4	1	0
101	100	12	2	24	4	1	2	31	9	5	3	0
102	mode	8.00	1.00	22.00	0.00	1.00	2.00	32.00	8.00	4.00	1.00	0.00
103	median	8.00	1.00	20.00	5.00			34.00	8.00	4.00	2.00	
104	mean	7.10		19.81	4.65			34.25	7.42			
105	min	-9.00	1.00	0.00	0.00			16.00	0.00	1.00	1.00	
106	max	22.00	3.00	46.00	12.00			51.00	15.00	5.00	4.00	
107	1st quartile	4.00	1.00	13.00	3.00			29.00	5.00	3.00	1.00	
108	3rd quartile	11.00	2.00	26.00	7.00			39.25	10.00	5.00	3.00	
109	st. dev.	5.53		9.68	3.39			7.62	3.51			
110	variance	30.54		93.65	11.50			58.09	12.35			
111		metric	ordinal	metric	metric	nominal	nominal	metric	metric	ordinal		ordinal	nominal

Average: 12.12627094   Count: 86   Sum: 800.3338824   Display Settings   100%

Chapter 5

5.1 Using the geometric mean, an average growth rate of rounded 5% is calculated for enterprise 1, and 7% for enterprise 5.

5.2 Charts help to communicate numbers and create lasting impressions; in addition, graphs often give clues about the behavior of people and objects or trends and patterns.

5.3 Frequency charts give a quick overview of the distribution of data, or how often individual values occur.

5.4 In the case of different class widths, the histogram is to be preferred. The histogram takes into account the different class widths.

5.5 Frequency table:

Expectation	Frequency, absolute	Frequency, relative	Frequency, cumulative
x_i	n_i	f_i	cf_i
1	2	0.33	0.33
2	3	0.50	0.83
3	1	0.17	1.00

Absolute and relative frequency plot:

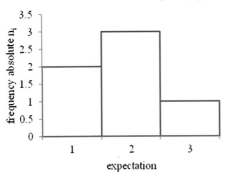
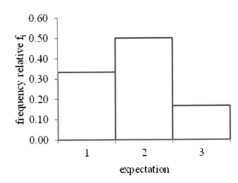

5.6 Frequency tables for marketing and innovation:

Marketing	Frequency, absolute	Frequency, relative	Innovation	Frequency, absolute	Frequency, relative
from over ... to ...	n_i	f_i	from over ... to ...	n_i	f_i
to 7	7	0.07	to 2	24	0.24
7 to 14	25	0.25	2 to 4	24	0.24
14 to 21	23	0.23	4 to 6	20	0.20
21 to 28	27	0.27	6 to 8	19	0.19
28 to 35	12	0.12	8 to 10	7	0.07
35 to 42	4	0.04	10 to 12	6	0.06
42 to 49	2	0.02	12 to 14	0	0.00

Other class widths are also possible.
Relative frequency charts:

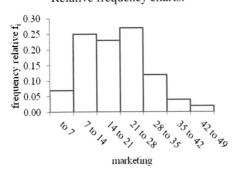
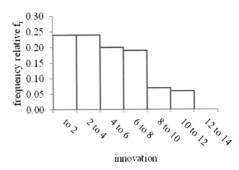

5.7 The charts from solution 5.6 show that neither the marketing variable nor the
innovation variable is truly symmetrical. Symmetry would exist if we could fold
each chart in the middle and have both sides come to rest on each other.

Growth rate	Frequency, absolute	Frequency, relative	Frequency density
from over ... to ...	n_i	f_i	f_i^*
-10 to -5	3	0.03	0.006
-5 to 0	8	0.08	0.016
0 to 5	27	0.27	0.054
5 to 25	62	0.62	0.031

5.8 Frequency table:

Relative frequency chart and histogram:

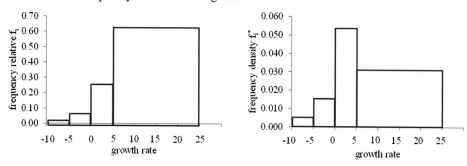

In the relative frequency chart, the right bar is much more dominant. This is due to the larger class width compared to the other classes. The histogram corrects for the different class widths and is therefore preferable.

5.9 Pie chart for the variable education

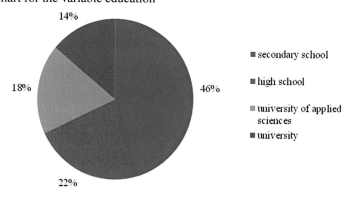

5.10 Bar chart for motives to start an enterprise by service and industrial firms

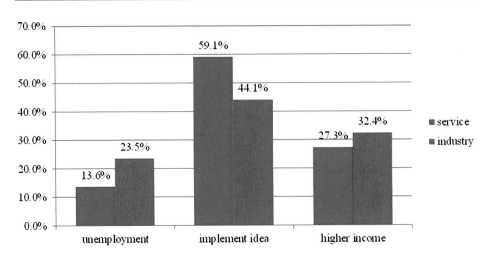

5.11 Line chart of the turnover development of the second enterprise in the data set

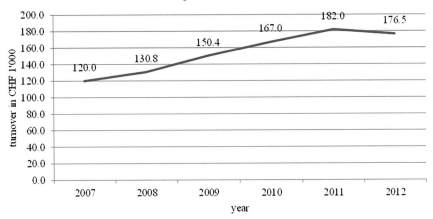

5.12 Development of overnight stays of selected Swiss holiday destinations (Source: BfS Switzerland, 2020, data downloaded from https://www.bfs.admin.ch/bfs/de/home/ statistics/tourism.html/ on 08.06.2020)

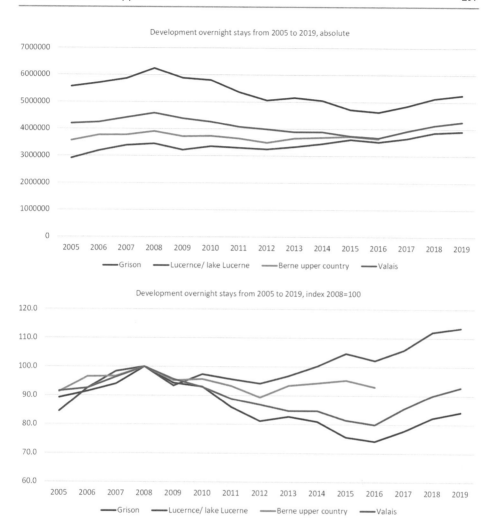

5.13 Population density in South America in 2018 (Source: World Bank, 2020, data downloaded from https://data.worldbank.org/ on 08/06/2020)

People per sq. km of land area, 2018

Unterstützt von Bing
© Microsoft, iomTom

Chapter 6

6.1 For nominal data with two characteristics, we use the phi coefficient. For nominal data with more than two characteristics, we use the contingency coefficient. For ordinal data, we use Spearman's correlation coefficient. For metric data, we use Bravais–Pearson's correlation coefficient.

6.2 Many correlations between two variables are spurious correlations. Whether there is a real correlation between two variables therefore requires always theoretical considerations.

6.3 A correlation coefficient only shows how two variables relate to each other. It does not tell whether one variable influences the other or vice versa. Statistical methods cannot do this. Causality is always a matter of theory.

6.4 In the figure on the left, we can approximately draw a straight line. Here the calculation of the Bravais–Pearson correlation coefficient is meaningful. The middle figure shows a perfect positive correlation, except for one outlier. Before we calculate the Bravais–Pearson correlation coefficient, we need to take care of the outlier. The figure on the right also shows a perfect, but non-linear correlation. Simply calculating the Bravais–Pearson correlation coefficient (without converting the correlation to a linear relationship) will provide incorrect results.

6.5 Scatterplot for the variables growth and marketing

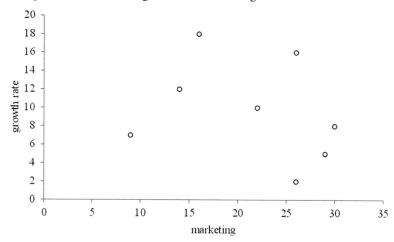

$r = -0.29$

6.6 Consider the respective scatter plots before calculation.

$$r_{growth,\,marketing} = 0.42$$

$$r_{growth,\,innovation} = 0.01$$

$$r_{growth,\,age} = 0.02$$

$$r_{growth,\,experience} = 0.62$$

6.7 $r_{sp} = 0.60$

6.8 $r_{sp} = 0.58$

6.9 $r_\phi = 0.13$

6.10 $C = 0.09$

6.11 Consider the scatterplot before the calculation. The calculation of the correlation coefficient according to Bravais–Pearson leads to a value of $r = -0.32$. But there is an outlier, without the outlier the value is $r = -0.79$. Depending on whether we calculate with outlier or without outlier, we obtain either a weak negative or a strong negative correlation between GDP per capita and CO_2-emissions per capita. Richer countries tend to emit less CO_2 per capita than poorer countries. Whether we calculate with or without outlier depends on whether the outlier is relevant for our analysis.

6.12 The correlation coefficient between the stocks of VW and Daimler is $r = 0.99$. Thus, we have an almost perfect positive relationship, i.e., the values goes exactly in the same direction. The correlation coefficient between the stocks of VW and SAP is $r = 0.56$ and that between Daimler and SAP $r = 0.48$, so we have a weaker positive correlation. We are interested in the correlation, for example, if we are aiming for risk diversification in our investment portfolio. Then the values of our stocks should not move exactly the same. However, there is also an outlier in the data here.

Chapter 7

7.1-3 See the examples in the book.

7.4 Measured in current prices, GDP per capita in China in 2018 is 363% of what it was in 2007, compared with 114% in Germany and 130% in Switzerland. Excluding inflation, this means that GDP per capita in China has grown the fastest, but from a much lower level.

7.5 If we calculate the Laspeyres price and quantity index, we see a 25% increase in prices from 2015 to 2019. This is partially offset by a reduction in volume of around 13%. If we calculate the price and quantity index according to Paasche, we see a price increase from 2015 to 2019 of 31% and a quantity reduction of about 9%.

7.6 At the base date of 6/10/2020, the index has moved as follows until 24/6/2020: 100.00 (10.6), 96.62, 93.73, 96.18, 98.94, 98.98, 99.20, 98.95, 98.18, 99.02, 95.54 (24.6). The average value of the portfolio is 4.46% ($95.54 - 100 = -4.46$) lower at 24.6 than at 10.6.

Chapter 8

8.1 First, the problem and the research question must be clearly defined. Second, assess what data are needed to answer the research question. Third, look for secondary data. If secondary data does not exist for the research question, we need to conduct a primary data survey.

8.2 Data can be found, for example, on the World Bank website.

8.3 The population are all persons or objects about which we want to make a statement. The sample is a part of the population. In order to infer from the sample to the population, we need an unbiased sample. We obtain such a sample if we draw the sample randomly.

8.4 $\bar{x}_{\text{growth}} = 7.1$, $\text{CI}_{90\%} = [6.19; 8.01]$, $\text{CI}_{95\%} = [6.01; 8.18]$, $\text{CI}_{99\%} = [5.67; 8.53]$

8.5 $n = \dfrac{2^2 \times 1.96^2 \times s^2}{\text{CIW}_{95\%}^2} = \dfrac{2^2 \times 1.96^2 \times 5^2}{2^2} = 96.04$.

8.6 $p_{\text{industry}} = 0.34$, $\text{CI}_{90\%} = [0.263; 0.417]$, $\text{CI}_{95\%} = [0.248; 0.432]$

$$\text{CI}_{99\%} = [0.219; 0.461]$$

8.7 $n = \dfrac{2^2 \times 1.64^2 \times p(1-p)}{\text{CIW}_{90\%}^2} = \dfrac{2^2 \times 1.64^2 \times 0.25(1-0.25)}{0.08^2} = 315.19$

8.8 Validity means that our variables actually measure the subject of the study. Reliability means that we measure it as precise as possible.

8.9 Statement one is correct, we can reliably measure the wrong thing. Statement two is not correct, if we do not measure reliably, then the variable is not a good proxy for the object of the study either.

Chapter 9

9.1 H_0: There is no relationship between turnover growth and the proportion of turnover spent on innovation.
H_A: There is a relationship between turnover growth and the proportion of turnover spent on innovation.
H_0: There is no difference between male and female founders in terms of the professional experience they have at the time of founding.
H_A: There is a difference between male and female founders in terms of the professional experience they have at the time of founding.
H_0: There is no relationship between smoking and life expectancy.
H_A: There is a relationship between smoking and life expectancy.

9.2 H_0: There is no correlation or a negative correlation between turnover growth and the proportion of turnover spent on innovation.
H_A: There is a positive correlation between turnover growth and the proportion of turnover spent on innovation.
H_0: Female founders have less or the same amount of professional experience than male founders at the time of founding.
H_A: Female founders have more professional experience than male founders at the time of founding.
H_0: Smoking has no or a positive influence on life expectancy.
H_A: Smoking has a negative influence on life expectancy.

9.5 No relationship and no difference are clearly defined statements and thus testable.

Chapter 10

10.1 Normal distributions:

10.2 $z_{(x=-25)} = -3$, $z_{(x=-20)} = -2.67$, $z_{(x=-10)} = -2$, $z_{(x=-7)} = -1.8$, $z_{(x=5)} = -1$, $z_{(x=20)} = 0$, $z_{(x=35)} = 1$, $z_{(x=47)} = 1.8$, $z_{(x=50)} = 2$, $z_{(x=60)} = 2.67$, $z_{(x=65)} = 3$

10.3 Area to the right of $z = 0.5$ is 30.85%, area to the right of $z = 0.75$ is 22.66%, area to the right of $z = 1.0$ is 15.87%, area to the right of $z = 1.26$ is 10.38%.

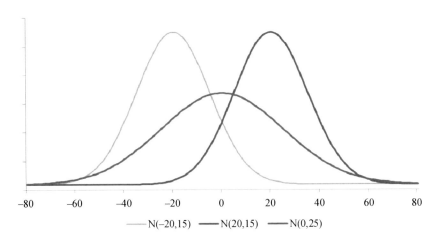

N(−20,15) ——— N(20,15) ——— N(0,25)

10.4 Area to the left of $z = -0.5$ is 30.85%, area to the left of $z = -0.75$ is 22.66%, area left of $z = -1.0$ is 15.87%, area left of $z = -1.26$ is 10.38%.

10.5 The areas to the right and left of the respective positive and negative z-values are equal due to the symmetry property of the standard normal distribution.

10.6 The probability of discovering a founder older than 42 years is 15.87%. The probability of discovering a founder younger than 21 years is 2.28%. The probability of discovering a founder within the interval of 42–49 years is 13.59%.

10.7 The probability is 6.58%.

10.8 To be among the top 5% of students, the score must exceed 96.5.

Chapter 11

11.1 The significance level α is, on the one hand, the probability at which we reject the null hypothesis, and on the other hand, it is the error probability of incorrectly rejecting the null hypothesis.

11.2 Statements:

- The statement that exceeding the critical value leads to the rejection of H_0 is correct.
- Even if you wanted to, you couldn't get the α-error to zero. The axes of the distribution function asymptotically approach the x-axis on, but never reach it.
- The statement that the smaller the α-error, the better the result is not correct. First, if the α-error gets smaller, we are less likely to reject the null hypothesis based on

our sample finding, even though this might be the correct decision. Second, there is not only an α-error but also a β-error, which increases with decreasing α.
- The statement that the choice of α depends on the effects of incorrectly rejecting the null hypothesis is correct.

11.3 Two-sided is tested when we have a non-directional null hypothesis. Left-sided or right-sided is tested when we have directional null hypothesis.

11.4 Two-sided test:
H_0: Enterprise founders are on average 40 years old.
H_A: Enterprise founders are on average not equal to 40 years old.
Left-sided test:
H_0: Enterprise founders are on average 40 years old or older.
H_A: Enterprise founders are on average younger than 40 years old.
Right-sided test:
H_0: Enterprise founders are on average 40 years old or younger.
H_A: Enterprise founders are on average older than 40 years.

11.5 The α-error is the probability to reject H_0, although H_0 is correct. The β-error is the probability to not reject H_0, although H_0 is not correct. If one reduces the α-error, the β-error becomes larger and vice versa.

Chapter 12

12.1 Step 1: Null hypothesis, alternative hypothesis, significance level

- H_0:$\mu = 10$
- H_A: $\mu \neq 10$
- $\alpha = 0.1$
  Step 2: Test distribution and test statistic
- Test distribution is the t-distribution with $df = n - 1 = 99$.
- Test statistic is $t = \frac{\bar{x} - \mu}{\hat{\sigma}_{\bar{x}}}$.
  Step 3: Rejection area and critical value

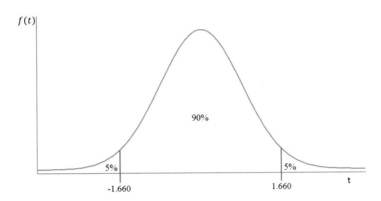

Step 4: Calculation of the test statistic

- $t = \frac{\bar{x} - \mu}{\hat{\sigma}_{\bar{x}}} = \frac{7.42 - 10}{\frac{3.51}{\sqrt{100}}} = -7.35$
- $Cohen's d = \frac{\bar{x} - \mu}{s} = \frac{7.42 - 10}{3.51} = -0.74$

Step 5: Decision and interpretation

- The calculated t-value is smaller than the critical t-value, we reject H_0.
- The professional experience is not 10 years, but is less. Cohen's d indicates a medium size effect.

12.2 Step 1: Null hypothesis, alternative hypothesis, significance level

- $H_0 : \mu = 5$
- $H_A : \mu \neq 5$
- $\alpha = 0.05$

  Step 2: Test distribution and test statistic
- Test distribution is the t-distribution with $df = n - 1 = 99$.
- Test statistic is $t = \frac{\bar{x} - \mu}{\hat{\sigma}_{\bar{x}}}$.

  Step 3: Rejection area and critical value

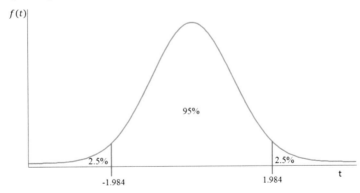

Step 4: Calculation of the test statistic
- $t = \frac{\bar{x}-\mu}{\hat{\sigma}_{\bar{x}}} = \frac{7.1-5}{\frac{5.53}{\sqrt{100}}} = 3.80$
- $Cohen's\, d = \frac{\bar{x}-\mu}{s} = \frac{7.1-5}{5.53} = 0.38$

Step 5: Decision and interpretation
- The calculated t-value is larger than the critical t-value, we reject H_0.
- The average firm growth is larger than 5%, thus our firm has grown below average. However, according to Cohen's d, the effect is rather small with 0.38 standard deviations.

12.3 Step 1: Null hypothesis, alternative hypothesis, significance level

- $H_0{:}\mu = 5$
- $H_A{:}\mu \neq 5$
- $\alpha = 0.05$

Step 2: Test distribution and test statistic
- Test distribution is the t-distribution with $df = n - 1 = 24$.
- Test statistic is $t = \frac{\bar{x}-\mu}{\hat{\sigma}_{\bar{x}}}$.

Step 3: Rejection area and critical value

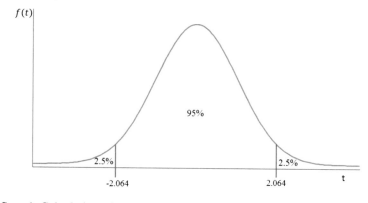

Step 4: Calculation of the test statistic
- $t = \frac{\bar{x}-\mu}{\hat{\sigma}_{\bar{x}}} = \frac{8-5}{\frac{4.16}{\sqrt{25}}} = 3.61$
- $Cohen's\, d = \frac{\bar{x}-\mu}{s} = \frac{8-5}{4.16} = 0.72$

Step 5: Decision and interpretation
- The calculated t-value is larger than the critical t-value, we reject H_0.
- The average firm growth is larger than 5%, thus our firm has grown below average. The effect is medium according to Cohen's d.

12.4 Step 1: Null hypothesis, alternative hypothesis, significance level.

- $H_0{:}\mu \geq 15$
- $H_A{:}\, \mu < 15$
- $\alpha = 0.1$
  Step 2: Test distribution and test statistic
- Test distribution is the t-distribution with $df = n - 1 = 99$.
- Test statistic is $t = \frac{\bar{x}-\mu}{\hat{\sigma}_{\bar{x}}}$.
  Step 3: Rejection area and critical value

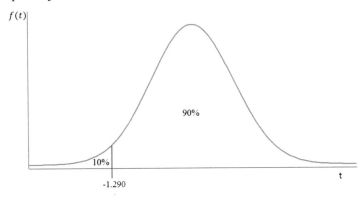

Step 4: Calculation of the test statistic
- $t = \frac{\bar{x}-\mu}{\hat{\sigma}_{\bar{x}}} = \frac{19.81-15}{\frac{9.677}{\sqrt{100}}} = 4.97$
- Since we do not reject H_0, we do not necessarily need to calculate the effect size.
  Step 5: Decision and interpretation
- The calculated t-value is larger than the critical t-value, we do not reject H_0.
- The proportion spent on marketing has not decreased.

12.5 Step 1: Null hypothesis, alternative hypothesis, significance level

- $H_0{:}\mu \leq 5$
- $H_A{:}\, \mu > 5$
- $\alpha = 0.05$
  Step 2: Test distribution and test statistic
- Test distribution is the t-distribution with $df = n - 1 = 99$.
- Test statistic is $t = \frac{\bar{x}-\mu}{\hat{\sigma}_{\bar{x}}}$.
  Step 3: Rejection area and critical value

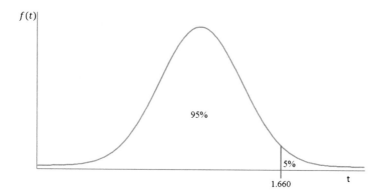

Step 4: Calculation of the test statistic

- $t = \frac{\bar{x}-\mu}{\hat{\sigma}_{\bar{x}}} = \frac{4.65-5}{\frac{3.392}{\sqrt{100}}} = -1.03$
- Since we do not reject H_0, we do not necessarily need to calculate the effect size.

Step 5: Decision and interpretation

- The calculated t-value is not larger than the critical t-value, we do not reject H_0.
- The proportion spent on innovation has not increased.

Chapter 13

13.1 Step 1: Null hypothesis, alternative hypothesis, significance level

- $H_0{:}\mu_M - \mu_F \le 0$
- $H_A{:}\mu_M - \mu_F > 0$
- $\alpha = 0.1$

Step 2: Test distribution and test statistic

- Test distribution is the t-distribution with $df = n_M + n_F - 2 = 98$.
- Test statistic is $t = \frac{\bar{x}_1 - \bar{x}_2}{\hat{\sigma}_{\bar{x}_1, \bar{x}_2}}$.

Step 3: Rejection area and critical value

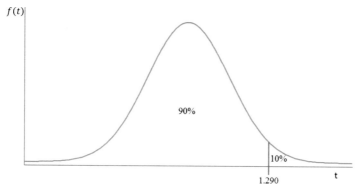

Step 4: Calculation of the test statistic

- $t = \frac{\bar{x}_1 - \bar{x}_2}{\hat{\sigma}_{\bar{x}_1, \bar{x}_2}} = \frac{34.77 - 33.29}{\sqrt{\left[\frac{(65-1)\times 7.55^2 + (35-1)\times 7.76^2}{65+35-2}\right]\left[\frac{65+35}{65\times 35}\right]}} = 0.93$

- The effect size would not necessarily need to be calculated, since we do not reject H_0. A calculation would yield the following value: $r = \sqrt{\frac{t^2}{t^2 + df}} = \sqrt{\frac{0.93^2}{0.93^2 + 98}} = 0.09$.

Step 5: Decision and interpretation

- The calculated t-value is not larger than the critical t-value, we do not reject H_0.
- Men are not older than women when starting an enterprise or there is no effect.

13.2-3 Step 1: Null hypothesis, alternative hypothesis, significance level

- $H_0: \mu_I - \mu_S = 0$
- $H_A: \mu_I - \mu_S \neq 0$
- $\alpha = 0.05$
  Steps 2, 3, and 4: Excel (calculation by hand is not shown)
- We test non-directional.
- We don't have to worry about the other steps, Excel calculates the relevant information.
- There are 34 industrial enterprises, the average growth rate is 6.35%.
- There are 66 service enterprises, the average growth rate is 7.48%.
- The variances are slightly different, we should consider running the test using unequal variances as well (we refrain from doing so here).
- The critical t-value for the two-sided test is ± 1.984.
- The calculated t value is -0.97.
- The effect size need not be calculated since we do not reject H_0.

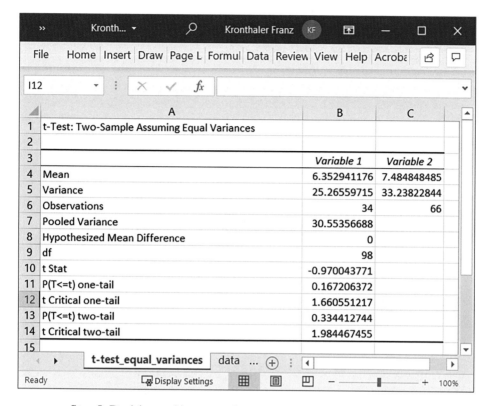

Step 5: Decision and interpretation
- The calculated t-value is not larger or smaller than the critical t-values, we do not reject H_0.
- There is no difference in the growth of industrial and service firms.

13.4-5 Step 1: Null hypothesis, alternative hypothesis, significance level

- H_0: $\mu_I - \mu_S \leq 0$
- H_A: $\mu_I - \mu_S > 0$
- $\alpha = 0.01$
  Steps 2, 3, and 4: Excel (calculation by hand is not shown)
- We test directional.
- Excel provides the other information.
- There are 34 industrial enterprises, the average expenditure on innovation is 4.71%.
- There are 66 service enterprises, the average spend on innovation is 4.62%.
- The variances are about the same.
- The critical t-value for the one-sided test is 2.365. We still need to consider whether the test is left-sided or right-sided. This depends on the null hypothesis

and the data entry. In our case we test right-sided and the critical t-value is +2.365.

- The calculated t-value is 0.12.
- We do not need to calculate the effect size because we do not reject H_0.

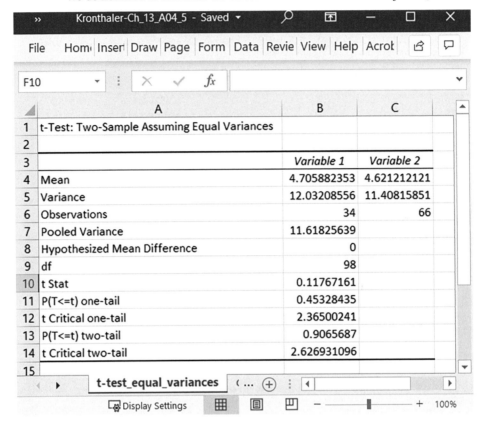

	A	B	C
1	t-Test: Two-Sample Assuming Equal Variances		
2			
3		*Variable 1*	*Variable 2*
4	Mean	4.705882353	4.621212121
5	Variance	12.03208556	11.40815851
6	Observations	34	66
7	Pooled Variance	11.61825639	
8	Hypothesized Mean Difference	0	
9	df	98	
10	t Stat	0.11767161	
11	P(T<=t) one-tail	0.45328435	
12	t Critical one-tail	2.36500241	
13	P(T<=t) two-tail	0.9065687	
14	t Critical two-tail	2.626931096	
15			

Step 5: Decision and interpretation
- The calculated t-value is not larger than the critical t-value, we do not reject H_0.
- Industrial firms do not spend more on innovation than service firms.

13.6-7 Step 1: Null hypothesis, alternative hypothesis, significance level

- $H_0: \mu_M - \mu_F = 0$
- $H_A: \mu_M - \mu_F \neq 0$
- $\alpha = 0.1$
  Steps 2, 3, and 4: Excel (calculation by hand is not shown)
- We test non-directional.
- Excel provides the other information.
- There are 65 founders, the average professional experience is 7.72 years.

- There are 35 female founders, the average professional experience is 6.86 years.
- The variances are about the same.
- The critical t-value in the two-sided test is ± 1.660.
- The calculated t-value is 1.18.
- The effect size need not be calculated since we do not reject H_0.

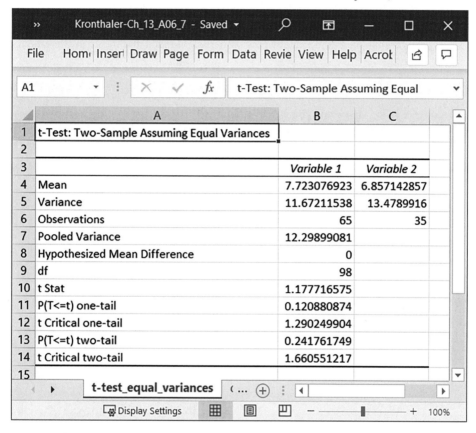

Step 5: Decision and interpretation
- The calculated t-value is not larger than the critical t-value, we do not reject H_0.
- Women and men are not different in terms of professional work experience.

13.8 Step 1: Null hypothesis, alternative hypothesis, significance level

- $H_0: \mu_M - \mu_F \leq 0$
- $H_A: \mu_M - \mu_F > 0$
- $\alpha = 0.1$ (is not specified in the task, may be set differently).

Step 2, 3 and 4: Excel

- We test directional.
- Excel provides the other information.
- There are 65 founders, the average expenditure on marketing is 20.26%.
- There are 35 female founders, the average expenditure on marketing is 18.97%.
- The variances are about the same.
- The critical t-value for the one-sided test is 1.290. We still need to consider whether the test is left-sided or right-sided. This depends on the null hypothesis and on the data entry. In our case we test right-sided and the critical t-value is +1.290.
- The calculated t-value is 0.63.
- We do not reject H_0, i.e. the effect size is not relevant.

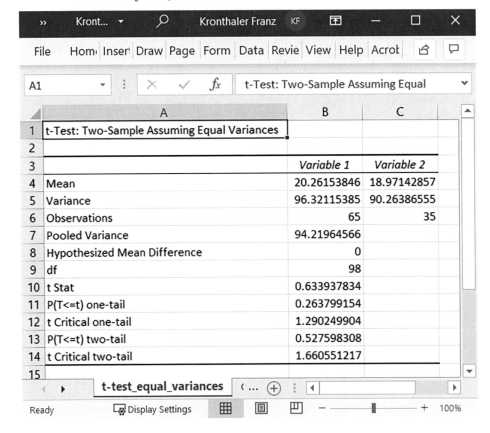

Step 5: Decision and interpretation
- The calculated t-value is not larger than the critical t-value, we do not reject H_0.
- Men do not spend more on marketing than women.

13.9 The arithmetic mean, standard deviation and sample size for both groups are needed.

13.10 The Mann-Whitney test can be used.

Chapter 14

14.1-2 Step 1: Null hypothesis, alternative hypothesis, significance level

- $H_0 : \mu_{Pre} - \mu_{Post} \leq 0$
- $H_A : \mu_{Pre} - \mu_{Post} > 0$
- $\alpha = 0.1$
  Step 2: Test distribution and test statistic
- Test distribution is the t-distribution with $df = n - 1 = 14$.
- Test statistic is $t = \dfrac{\sum d_i}{\sqrt{\frac{n \sum d_i^2 - (\sum d_i)^2}{n-1}}}$.

  Step 3: Rejection area and critical value

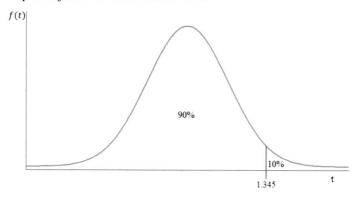

Step 4: Calculation of the test statistic
- $t = \dfrac{\sum d_i}{\sqrt{\frac{n \sum d_i^2 - (\sum d_i)^2}{n-1}}} = \dfrac{17}{\sqrt{\frac{15 \times 161 - 289}{15-1}}} = 1.38$
- effect size $r = \sqrt{\dfrac{t^2}{t^2 + Fg}} = \sqrt{\dfrac{1.38^2}{1.38^2 + 14}} = 0.35$
  Step 5: Decision and interpretation
- The calculated t-value is larger than the critical t-value, we reject H_0.
- The training works and achieves a medium effect.
  The following table shows the Excel results:

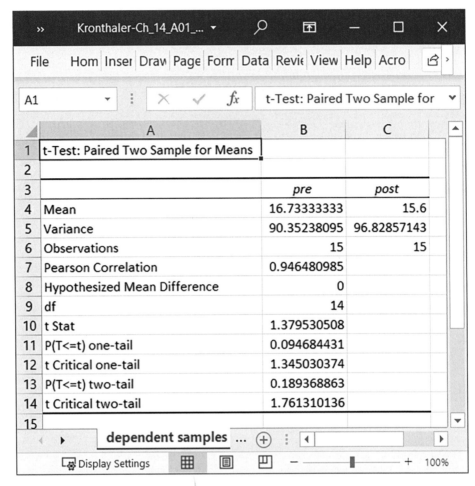

14.3-4 Step 1: Null hypothesis, alternative hypothesis, significance level

- $H_0: \mu_{without} - \mu_{with} \geq 0$
- $H_A: \mu_{without} - \mu_{with} < 0$
- $\alpha = 0.01$ (14.3) and $\alpha = 0.1$ (14.4)
  Steps 2, 3, and 4: Excel (calculation by hand is not shown)
- We test directional.
- Excel provides the other information.
- The average distance run without energy drinks was 15.6 km.
- The average distance run with energy drinks was 16.2 km.
- The critical t-value for the one-sided test is 2.821. We still have to decide whether the test is performed on the left or the right side. This depends on the null hypothesis. In our case we test left-sided and the critical t-value is -2.821.
- The calculated t-value is -1.20.
- The effect size does not need to be calculated because H_0 is not rejected.

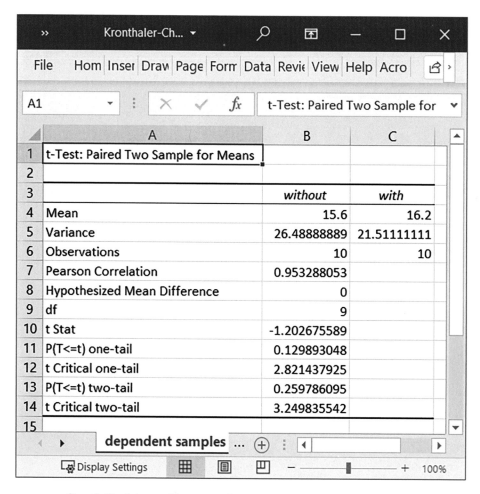

Step 5: Decision and interpretation
- The calculated t-value is larger than the critical t-value, we do not reject H_0.
- Energy drinks do not increase the performance.

Chapter 15

15.1 Step 1: Null hypothesis, alternative hypothesis, significance level

- H_0: $\mu_1 = \mu_2 = \mu_3$
- H_A: $\mu_i \neq \mu_j$ for at least for one pair of ij
- $\alpha = 0.1$
  Step 2: Test distribution and test statistic
- Test distribution is the F-distribution with 2 and 9 degrees of freedom.

- Test statistic is

$$F = \frac{SS_{explained}/(G-1)}{SS_{not-explained}/(G \times (K-1))}$$

Step 3: Rejection area and critical value
- The test distbribution is the F-distribution with $\alpha = 10\%$ and $df_1 = 2$ and $df_2 = 9$.
- The critical F-value is 3.01.
  Step 4: Calculate test statistic

AutoSave ● Off   🔲 ⤾ ⤿ 🗐 ⤍   ⤶ Kronthal... - Saved ▾   🔎   Kronthaler Franz   KF   ⊞   —   ☐   ✕

File   Home   Insert   Draw   Page Layout   Formulas   Data   Review   View   Help   Acrobat   🖆 Share   ☐

A1   ▾   :   ✕   ✓   *fx*

	A	B	C	D	E	F	G
1					variance total	variance explained	variance unexplained
2	enterprise	workload	motive	group means	$(y_{gk}\text{-}y_{mw})^2$	$(y_{g,mw}\text{-}y_{mw})^2$	$(y_{gk}\text{-}y_{g,mw})^2$
3	1	37	1		230.03	119.17	18.06
4	2	40	1		148.03	119.17	1.56
5	5	43	1		84.03	119.17	3.06
6	12	45	1	41.25	51.36	119.17	14.06
7	3	60	2		61.36	40.11	2.25
8	4	62	2		96.69	40.11	12.25
9	9	57	2		23.36	40.11	2.25
10	10	55	2	58.5	8.03	40.11	12.25
11	6	54	3		3.36	21.01	7.56
12	7	58	3		34.03	21.01	1.56
13	8	60	3		61.36	21.01	10.56
14	11	55	3	56.75	8.03	21.01	3.06
15			global mean	52.17	809.67	721.17	88.50
16				df	G*K-1	G-1	G*(K-1)
17					11	2	9
18					mss total	mss explained	mss unexplained
19					73.61	360.58	9.83
20				F	36.67		
21	Critical F-value, F-distribution alpha 10 % with 2 and 9 degrees of freedom				3.01		

◂ ▸ │ data │ legend │ **solution** │ ⊕

🖳 Display Settings   ⊞ ▣ ◳ — ⟍—|⟍— + 100%

- $F = \frac{SS_{explained}/(G-1)}{SS_{unexplained}/(G \times (K-1))} = \frac{721.17/(3-1)}{88.50/(3 \times (4-1))} = 36.67$
- effect size $r = \sqrt{\frac{SS_{explained}}{SS_{total}}} = \sqrt{\frac{721.17}{809.67}} = 0.94$
  Step 5: Decision and interpretation
- The calculated F-value is larger than the critical F-value, we reject H_0.
- The effect size indicates a large effect, this is also reflected in the absolute numbers.
- There is a difference in working time at least between two groups, to determine between which, the post hoc test has to be conducted.

15.2 Step 1: Null hypothesis, alternative hypothesis, significance level

- $H_0\colon \mu_1 = \mu_2 = \mu_3 = \mu_4$
- $H_A\colon \mu_i \neq \mu_j$ for at least for one pair of ij
- $\alpha = 0.05$

  Step 2, 3 and 4: Excel
- The test statistic is the F-value.
- The test distribution is the F-distribution with 3 and 96 degrees of freedom
- We take the other information from the Excel output.
- We do not need to calculate the effect size because we do not reject H_0.

Anova: Single Factor					
SUMMARY					
Groups	*Count*	*Sum*	*Average*	*Variance*	
secondary school	46	308	6.70	34.66	
high school	22	166	7.55	26.64	
university of applied sciences	18	123	6.83	38.97	
university	14	113	8.07	16.53	
ANOVA					
Source of Variation	*SS*	*df*	*MS*	*F*	*P-value*	*F crit*
Between Groups	26.38	3	8.79	0.28	0.84	2.70
Within Groups	2996.62	96	31.21			
Total	3023	99			

Step 5: Decision and interpretation
- The calculated F-value does not exceed the critical F-value, we do not reject H_0.
- The result indicates that there is no difference in business growth among the founders with different levels of education.

15.3 Step 1: Null hypothesis, alternative hypothesis, significance level

- $H_0\colon \mu_1 = \mu_2 = \mu_3 = \mu_4$
- $H_A\colon \mu_i \neq \mu_j$ for at least one pair of ij
- $\alpha = 0.1$

Step 2, 3 and 4: Excel
- The test statistic is the F-value.
- The test distribution is the F-distribution with 3 and 96 degrees of freedom.
- We take the other information from the Excel output.
- We do not need to calculate the effect size because we do not reject H_0.

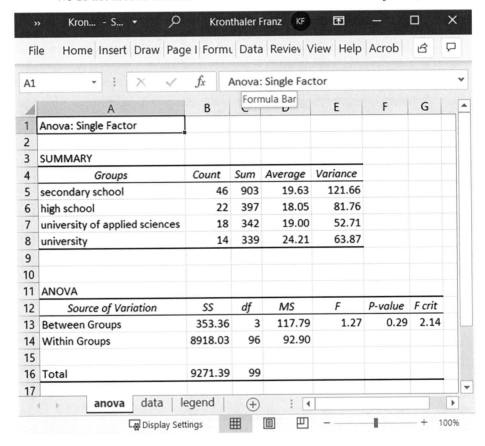

Step 5: Decision and interpretation
- The calculated F-value does not exceed the critical F-value, we do not reject H_0.
- The result indicates that there is no difference in marketing expenditure among founders with different educational background.

15.4 Step 1: Null hypothesis, alternative hypothesis, significance level

- H_0: $\mu_1 = \mu_2$
- H_A: $\mu_1 \neq \mu_2$
- $\alpha = 0.05$
  Step 2, 3, 4 and 5: Independent samples t-test and analysis of variance using Excel

- We do not exceed the critical value in both cases and do not reject H_0.
- There is no difference in weekly working hours between female and male founders.
- The analysis of variance and the independent samples t-test, provide the same result.

Chapter 16

16.1 Step 1: Null hypothesis, alternative hypothesis, significance level

- $H_0: r = 0$
- $H_A: r \neq 0$
- $\alpha = 0.05$
    Step 2: Test distribution and test statistic
- Test distribution is the t-distribution with $df = 8 - 2 = 6$.
- Test statistic is $t = \frac{r \times \sqrt{n-2}}{\sqrt{1-r^2}}$.
    Step 3: Rejection area and critical value

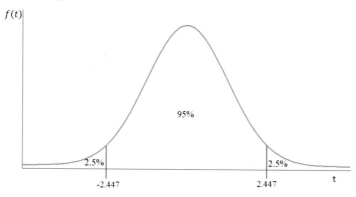

    Step 4: Calculation of the test statistic
- First, we need to look at the scatterplot.
- Then we can calculate the Bravais–Pearson correlation coefficient. It is $r = -0.29$.
- $t = \frac{-0.29 \times \sqrt{8-2}}{\sqrt{1-(-0.29)^2}} = -0.74$
    Step 5: Decision and interpretation
- The calculated t-value is not larger or smaller than the critical t-value, we do not reject H_0.
- There is no relationship between the two variables.

16.2 Step 1: Null hypothesis, alternative hypothesis, significance level

- $H_0: r = 0$
- $H_A: r \neq 0$
- $\alpha = 0.05$
    Step 2: Test distribution and test statistic
- Test distribution is the t-distribution with $df = 100 - 2 = 98$.
- Test statistic is $t = \frac{r \times \sqrt{n-2}}{\sqrt{1-r^2}}$.
    Step 3: Rejection area and critical value
- The critical t-value is ± 1.984.
- We reject H_0 if we fall below -1.984 or exceed 1.984.
    Step 4: Calculation of the test statistic
- First, we need to look at the scatterplot.
- Then we can calculate the Bravais–Pearson correlation coefficient using Excel. It is $r = 0.42$.
- $t = \frac{0.42 \times \sqrt{100-2}}{\sqrt{1-0.42^2}} = 4.581$
    Step 5: Decision and interpretation
- The calculated t-value is larger than the critical t-value, we reject H_0.
- There is a medium relationship between the two variables.

16.3 Step 1: Null hypothesis, alternative hypothesis, significance level

- $H_0: r \leq 0$
- $H_A: r > 0$
- $\alpha = 0.01$
    Step 2: Test distribution and test statistic
- Test distribution is the t-distribution with $df = n - 2 = 8 - 6 = 2$.
- Test statistic is $t = \frac{r \times \sqrt{n-2}}{\sqrt{1-r^2}}$.
    Step 3: Rejection area and critical value

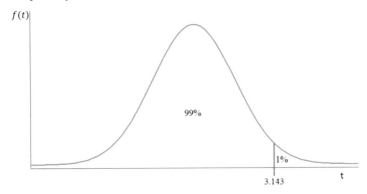

Step 4: Calculation of the test statistic
- First, we need to look at the scatterplot.
- Then we can determine the Bravais–Pearson correlation coefficient. It is $r = 0.90$.
- $t = \frac{0.90 \times \sqrt{8-2}}{\sqrt{1-0.90^2}} = 5.06$

Step 5: Decision and interpretation
- The calculated t-value is larger than the critical t-value, we reject H_0.
- There is a strong positive relationship between the two variables.

16.4 Step 1: Null hypothesis, alternative hypothesis, significance level

- $H_0: r \leq 0$
- $H_A: r > 0$
- $\alpha = 0.01$

Step 2: Test distribution and test statistic
- Test distribution is the t-distribution with $df = n - 2 = 100 - 2 = 98$.
- Test statistic is $t = \frac{r \times \sqrt{n-2}}{\sqrt{1-r^2}}$.

Step 3: Rejection area and critical value
- The critical t-value is 2.364.
- We reject H_0 if we exceed 2.364.

Step 4: Calculation of the test statistic
- First, we need to look at the scatterplot.
- Then we can calculate the Bravais–Pearson correlation coefficient using Excel. It is $r = 0.62$.
- $t = \frac{0.62 \times \sqrt{100-2}}{\sqrt{1-0.62^2}} = 7.823$

Step 5: Decision and interpretation
- The calculated t-value is larger than the critical t-value, we reject H_0.
- There is a strong positive relationship between the two variables.

16.5 Step 1: Null hypothesis, alternative hypothesis, significance level

- $H_0: r_{Sp} \leq 0$
- $H_A: r_{Sp} > 0$
- $\alpha = 0.1$

Step 2: Test distribution and test statistic
- Test distribution is the t-distribution with $df = n - 2 = 8 - 2 = 6$.
- Test statistic is $t = \frac{r_{Sp}}{\sqrt{\frac{1-r_{Sp}^2}{n-2}}}$.

Step 3: Rejection area and critical value

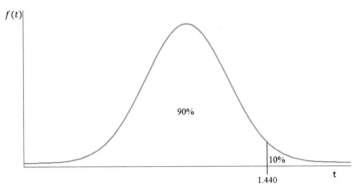

Step 4: Calculation of the test statistic
- We calculate Spearman's correlation coefficient. It is $r_{Sp} = 0.57$.
- $t = \dfrac{0.57}{\sqrt{\frac{1-0.57^2}{8-2}}} = 1.70$

Step 5: Decision and interpretation
- The calculated t-value is larger than the critical t-value, we reject H_0.
- There is a moderate positive relationship between the two variables.

16.6 Step 1: Null hypothesis, alternative hypothesis, significance level

- H_0: $r_{Sp} \leq 0$
- H_A: $r_{Sp} > 0$
- $\alpha = 0.1$

Step 2: Test distribution and test statistic
- Test distribution is the t-distribution with $df = n - 2 = 100 - 2 = 98$.
- Test statistic is $t = \dfrac{r_{Sp}}{\sqrt{\frac{1-r_{Sp}^2}{n-2}}}$.

Step 3: Rejection area and critical value
- The critical t-value is 1.290.
- We reject H_0 if we exceed 1.290.

Step 4: Calculation of the test statistic
- We calculate Spearman's correlation coefficient. It is $r_{Sp} = 0.58$.
- $t = \dfrac{0.58}{\sqrt{\frac{1-0.58^2}{100-2}}} = 7.05$

Step 5: Decision and interpretation
- The calculated t-value is larger than the critical t-value, we reject H_0.
- There is a moderate positive relationship between the two variables.

16.7 Step 1: Null hypothesis, alternative hypothesis, significance level

- H_0: $r_\Phi = 0$
- H_A: $r_\Phi \neq 0$
- $\alpha = 0.05$
    Step 2: Test distribution and test statistic

- Test distribution is the $\chi 2$ distribution with one degree of freedom.
- Test statistic is $\chi^2 = n \times r_\Phi^2$.
    Step 3: Rejection area and critical value

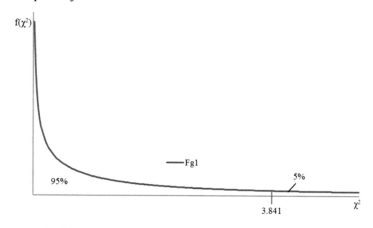

    Step 4: Calculation of the test statistic
- We calculate the phi coefficient. It is $r_\Phi = 0.13$.
- $\chi^2 = n \times r_\Phi^2 = 100 \times 0.13^2 = 1.69$
    Step 5: Decision and interpretation
- The calculated $\chi 2$-value is not larger than the critical value, we do not reject H_0.
- There is no relationship between the two variables.

16.8 Step 1: Null hypothesis, alternative hypothesis, significance level

- H_0: $C = 0$
- H_A: $C > 0$
- $\alpha = 0.1$
    Step 2: Test distribution and test statistic
- Test distribution is the $\chi 2$-distribution with $(k - 1) \times (j - 1) = 1 \times 2 = 2$ degrees of freedom.

- Test statistic is $U = \sum \sum \frac{(f_{jk}-e_{jk})^2}{e_{jk}}$.
  Step 3: Rejection area and critical value

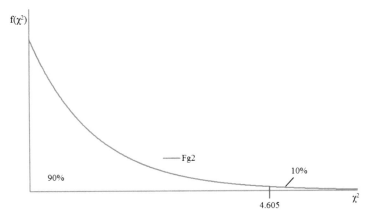

Step 4: Calculation of the test statistic
- The contingency coefficient is $C = 0.09$ (for the test, however, we only need U).
- $U = 0.793$
  Step 5: Decision and interpretation
- The calculated U-value is not larger than the critical value, we do not reject H_0.
- There is no relationship between the two variables.

16.9 There is a significant relationship only between the shares of Daimler and VW, i.e. based on the data we can assume that the value of the two stocks move in the same direction.

Chapter 17

17.1 Step 1: Null hypothesis, alternative hypothesis, significance level

- H_0: $\pi_1 = \pi_2 = \frac{1}{3}$
- H_A: $\pi_1 \neq \pi_2 \neq \pi_2 \neq \frac{1}{3}$
- $\alpha = 0.05$
  Step 2: Test distribution and test statistic
- Test distribution is the χ^2 distribution with. $df = c - 1 = 2$.
- Test statistic is $U = \sum \frac{(f_i - e_i)^2}{e_i}$.
  Step 3: Rejection area and critical value

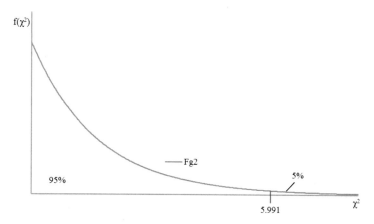

Step 4: Calculation of the test statistic
- $U = \sum \frac{(f_i - e_i)^2}{e_i} = \frac{(17-33.3)^2}{33.3} + \frac{(54-33.3)^2}{33.3} + \frac{(29-33.3)^2}{33.3} = 21.40$

Step 5: Decision and interpretation
- The value of the test statistic is larger than the critical value, we reject H_0.
- The founding motives are not equally frequent.

17.2 Step 1: Null hypothesis, alternative hypothesis, significance level

- H_0: $\pi_{women} = \pi_{men}$
- H_A: $\pi_{women} \neq \pi_{men}$
- $\alpha = 0.05$

Step 2: Test distribution and test statistic
- Test distribution is the $\chi2$ distribution with $df = (j-1) \times (k-1) = 2$.
- Test statistic is $U = \sum \sum \frac{(f_{jk} - e_{jk})^2}{e_{jk}}$.

Step 3: Rejection area and critical value

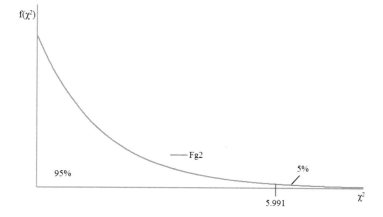

Step 4: Calculation of the test statistic
- $U = 0.793$

Step 5: Decision and interpretation
- The calculated test statistic is not larger than the critical value, we do not reject H_0.
- There are no differences between men and women in terms of motives for founding.

17.3 Step 1: Null hypothesis, alternative hypothesis, significance level

- $H_0: \pi_{before} = \pi_{after}$
- $H_A: \pi_{vor} \neq \pi_{after}$
- $\alpha = 0.1$

Step 2: Test distribution and test statistic
- Test distribution is the $\chi 2$ distribution with one degree of freedom.
- Test statistic is $U = \sum \frac{(f_i - e_i)^2}{e_i}$.

Step 3: Rejection area and critical value

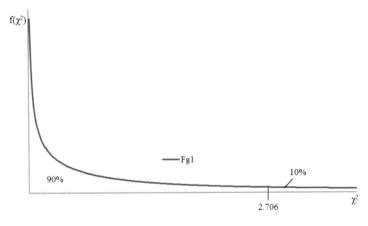

Step 4: Calculation of the test statistic

-

$$U = \sum \frac{(f_i - e_i)^2}{e_i} = \frac{(f_b - e_b)^2}{e_b} + \frac{(f_c - e_c)^2}{e_c} = \frac{(40 - 55)^2}{55} + \frac{(70 - 55)^2}{55} = 8.18$$

Step 5: Decision and interpretation

- The calculated test statistic is larger than the critical value, we reject H_0.
- The educational campaign has an impact on dietary behavior.

Chapter 19

19.1 Without theory about the relationship, the result of the regression analysis is worthless. Only with the help of theory can we make a statement about whether one variable affects another variable. If we do not have this statement, we do not know what will happen to Y if we change X.

19.2 Result:

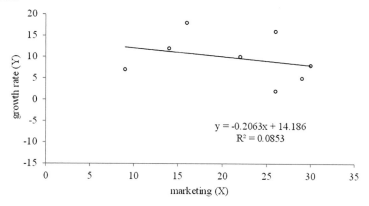

- The slope of the regression line is slightly negative, tending to decrease the growth rate when marketing increases.
- The R^2 is only 8.5%, i.e. the explanatory power of the regression line is very low.

19.3 The predicted growth rate is 10.06%. The forecast is very unreliable because the coefficient of determination is R^2 is very low (see task 19.2).

19.4 Result:

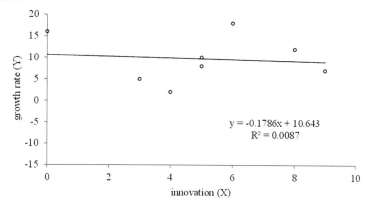

- The regression line is $\hat{y} = 10.643 - 0.1786x$. The coefficient of determination is 0.87%.
- The R^2 is only 0.87%, i.e. the explanatory power is close to zero.

19.5 The predicted growth rate is 7.07%. We have two problems with the forecast: (1) it is an out-of-sample forecast, (2) the coefficient of determination R^2 is very small (see task 19.4).

19.6 Result:

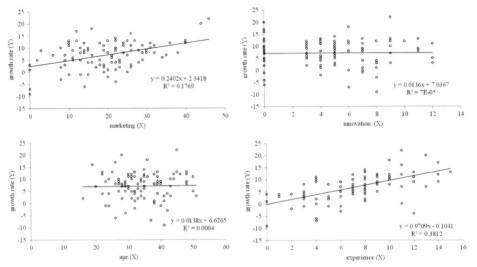

- There seems to be a relationship between growth and marketing and growth and experience.
- There seems to be no relationship between the growth and innovation and growth and age.

19.7 Both theoretically and in view of task 19.6, it is very likely that not only one variable but several variable have an influence on the growth rate.

Chapter 20

20.1 Result:

	A	B	C	D	E	F	G
1	SUMMARY OUTPUT						
2							
3	Regression Statistics						
4	Multiple R	0.69					
5	R Square	0.48					
6	Adjusted R Square	0.45					
7	Standard Error	4.08					
8	Observations	100					
9							
10	ANOVA						
11		df	SS	MS	F	Significance F	
12	Regression	5	1456.06	291.21	17.47	0.00	
13	Residual	94	1566.94	16.67			
14	Total	99	3023				
15							
16		Coefficients	Standard Error	t Stat	P-value	Lower 95%	Upper 95%
17	Intercept	-4.18	1.53	-2.74	0.01	-7.21	-1.15
18	marketing	0.15	0.04	3.34	0.00	0.06	0.24
19	innovation	0.28	0.13	2.23	0.03	0.03	0.53
20	sector	0.33	0.89	0.37	0.71	-1.44	2.09
21	experience	0.93	0.13	7.32	0.00	0.68	1.18
22	sex	-0.28	0.87	-0.32	0.75	-2.01	1.46

- For the interpretation compare Chap. 20.

20.2 Result Linearity:

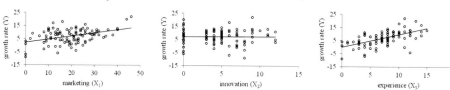

- The scatterplots show either a linear or no relationship between the dependent and independent variables.

- No non-linear relationship can be detected.
  Correlation of the residuals with the independent variables:

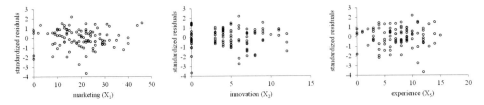

- No relationship can be detected between the independent variables and the residuals.
  Constancy of variance of the residuals:

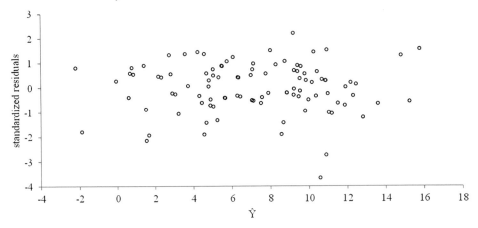

- The deviation of the residuals does not change with an increase or an decrease of the estimated $\hat{Y}$-values (the residuals scatter largely in the same range).

No correlation between two or more independent variables:

	Marketing	Innovation	Experience
Marketing	1		
Innovation	-0.03	1	
Experience	0.26	-0.27	1

- There is no correlation between the independent variables.

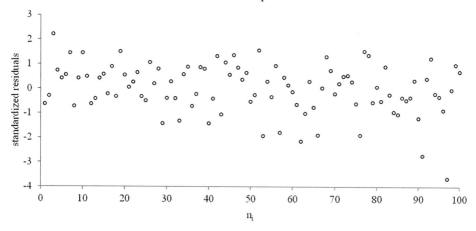

We do not detect any pattern in the figure.
Normal distribution of residuals:

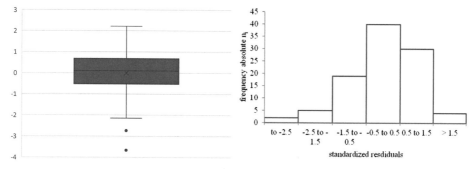

- The median is approximately in the middle of the box, i.e. the variable seems to be symmetrical. The histogram shows a slightly left-skewed distribution.
- One more note here about the boxplot, two observations have very large standardized deviation and are marked as outliers. Outliers tend to have an impact on regression results and the respective observations should be considered more closely.

20.3 The calculation of the correlations shows that there is no longer a high correlation between the independent variables. This is an indication that there is no problem with the assumption of multicollinearity.

The Standard Normal Distribution N (0,1)

B

© Springer-Verlag GmbH Germany, part of Springer Nature 2023
F. Kronthaler, *Statistics Applied With Excel*,
https://doi.org/10.1007/978-3-662-64319-8

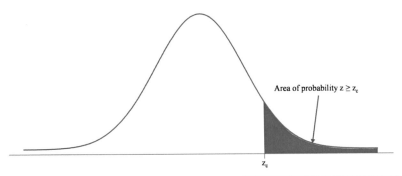

Area of probability $z \geq z_c$

z_c

z_k →					second decimal place					
↓	.00	.01	.02	.03	.04	.05	.06	.07	.08	.09
0.0	0.5000	0.4960	0.4920	0.4880	0.4840	0.4801	0.4761	0.4721	0.4681	0.4641
0.1	0.4602	0.4562	0.4522	0.4483	0.4443	0.4404	0.4364	0.4325	0.4286	0.4247
0.2	0.4207	0.4168	0.4129	0.4090	0.4052	0.4013	0.3974	0.3936	0.3897	0.3859
0.3	0.3821	0.3783	0.3745	0.3707	0.3669	0.3632	0.3594	0.3557	0.3520	0.3483
0.4	0.3446	0.3409	0.3372	0.3336	0.3300	0.3264	0.3228	0.3192	0.3156	0.3121
0.5	0.3085	0.3050	0.3015	0.2981	0.2946	0.2912	0.2877	0.2843	0.2810	0.2776
0.6	0.2743	0.2709	0.2676	0.2643	0.2611	0.2578	0.2546	0.2514	0.2483	0.2451
0.7	0.2420	0.2389	0.2358	0.2327	0.2296	0.2266	0.2236	0.2206	0.2177	0.2148
0.8	0.2119	0.2090	0.2061	0.2033	0.2005	0.1977	0.1949	0.1922	0.1894	0.1867
0.9	0.1841	0.1814	0.1788	0.1762	0.1736	0.1711	0.1685	0.1660	0.1635	0.1611
1.0	0.1587	0.1562	0.1539	0.1515	0.1492	0.1469	0.1446	0.1423	0.1401	0.1379
1.1	0.1357	0.1335	0.1314	0.1292	0.1271	0.1251	0.1230	0.1210	0.1190	0.1170
1.2	0.1151	0.1131	0.1112	0.1093	0.1075	0.1056	0.1038	0.1020	0.1003	0.0985
1.3	0.0968	0.0951	0.0934	0.0918	0.0901	0.0885	0.0869	0.0853	0.0838	0.0823
1.4	0.0808	0.0793	0.0778	0.0764	0.0749	0.0735	0.0721	0.0708	0.0694	0.0681
1.5	0.0668	0.0655	0.0643	0.0630	0.0618	0.0606	0.0594	0.0582	0.0571	0.0559
1.6	0.0548	0.0537	0.0526	0.0516	0.0505	0.0495	0.0485	0.0475	0.0465	0.0455
1.7	0.0446	0.0436	0.0427	0.0418	0.0409	0.0401	0.0392	0.0384	0.0375	0.0367
1.8	0.0359	0.0351	0.0344	0.0336	0.0329	0.0322	0.0314	0.0307	0.0301	0.0294
1.9	0.0287	0.0281	0.0274	0.0268	0.0262	0.0256	0.0250	0.0244	0.0239	0.0233
2.0	0.0228	0.0222	0.0217	0.0212	0.0207	0.0202	0.0197	0.0192	0.0188	0.0183
2.1	0.0179	0.0174	0.0170	0.0166	0.0162	0.0158	0.0154	0.0150	0.0146	0.0143
2.2	0.0139	0.0136	0.0132	0.0129	0.0125	0.0122	0.0119	0.0116	0.0113	0.0110
2.3	0.0107	0.0104	0.0102	0.0099	0.0096	0.0094	0.0091	0.0089	0.0087	0.0084
2.4	0.0082	0.0080	0.0078	0.0075	0.0073	0.0071	0.0069	0.0068	0.0066	0.0064
2.5	0.0062	0.0060	0.0059	0.0057	0.0055	0.0054	0.0052	0.0051	0.0049	0.0048
2.6	0.0047	0.0045	0.0044	0.0043	0.0041	0.0040	0.0039	0.0038	0.0037	0.0036
2.7	0.0035	0.0034	0.0033	0.0032	0.0031	0.0030	0.0029	0.0028	0.0027	0.0026
2.8	0.0026	0.0025	0.0024	0.0023	0.0023	0.0022	0.0021	0.0021	0.0020	0.0019
2.9	0.0019	0.0018	0.0018	0.0017	0.0016	0.0016	0.0015	0.0015	0.0014	0.0014
3.0	0.001350									
3.5	0.000233									
4.0	0.000032									
5.0	0.0000003									

The t-Distribution

© Springer-Verlag GmbH Germany, part of Springer Nature 2023
F. Kronthaler, *Statistics Applied With Excel*,
https://doi.org/10.1007/978-3-662-64319-8

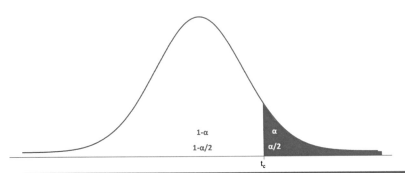

df	0.200	0.150	0.125	0.100	α or α/2 0.075	0.050	0.025	0.010	0.005
1	1.376	1.963	2.414	3.078	4.165	6.314	12.706	31.821	63.657
2	1.061	1.386	1.604	1.886	2.282	2.920	4.303	6.965	9.925
3	0.978	1.250	1.423	1.638	1.924	2.353	3.182	4.541	5.841
4	0.941	1.190	1.344	1.533	1.778	2.132	2.776	3.747	4.604
5	0.920	1.156	1.301	1.476	1.699	2.015	2.571	3.365	4.032
6	0.906	1.134	1.273	1.440	1.650	1.943	2.447	3.143	3.707
7	0.896	1.119	1.254	1.415	1.617	1.895	2.365	2.998	3.499
8	0.889	1.108	1.240	1.397	1.592	1.860	2.306	2.896	3.355
9	0.883	1.100	1.230	1.383	1.574	1.833	2.262	2.821	3.250
10	0.879	1.093	1.221	1.372	1.559	1.812	2.228	2.764	3.169
11	0.876	1.088	1.214	1.363	1.548	1.796	2.201	2.718	3.106
12	0.873	1.083	1.209	1.356	1.538	1.782	2.179	2.681	3.055
13	0.870	1.079	1.204	1.350	1.530	1.771	2.160	2.650	3.012
14	0.868	1.076	1.200	1.345	1.523	1.761	2.145	2.624	2.977
15	0.866	1.074	1.197	1.341	1.517	1.753	2.131	2.602	2.947
16	0.865	1.071	1.194	1.337	1.512	1.746	2.120	2.583	2.921
17	0.863	1.069	1.191	1.333	1.508	1.740	2.110	2.567	2.898
18	0.862	1.067	1.189	1.330	1.504	1.734	2.101	2.552	2.878
19	0.861	1.066	1.187	1.328	1.500	1.729	2.093	2.539	2.861
20	0.860	1.064	1.185	1.325	1.497	1.725	2.086	2.528	2.845
21	0.859	1.063	1.183	1.323	1.494	1.721	2.080	2.518	2.831
22	0.858	1.061	1.182	1.321	1.492	1.717	2.074	2.508	2.819
23	0.858	1.060	1.180	1.319	1.489	1.714	2.069	2.500	2.807
24	0.857	1.059	1.179	1.318	1.487	1.711	2.064	2.492	2.797
25	0.856	1.058	1.178	1.316	1.485	1.708	2.060	2.485	2.787
26	0.856	1.058	1.177	1.315	1.483	1.706	2.056	2.479	2.779
27	0.855	1.057	1.176	1.314	1.482	1.703	2.052	2.473	2.771
28	0.855	1.056	1.175	1.313	1.480	1.701	2.048	2.467	2.763
29	0.854	1.055	1.174	1.311	1.479	1.699	2.045	2.462	2.756
30	0.854	1.055	1.173	1.310	1.477	1.697	2.042	2.457	2.750
40	0.851	1.050	1.167	1.303	1.468	1.684	2.021	2.423	2.704
50	0.849	1.047	1.164	1.299	1.462	1.676	2.009	2.403	2.678
100	0.845	1.042	1.157	1.290	1.451	1.660	1.984	2.364	2.626
200	0.843	1.039	1.154	1.286	1.445	1.653	1.972	2.345	2.601

The χ^2-Distribution

D

© Springer-Verlag GmbH Germany, part of Springer Nature 2023
F. Kronthaler, *Statistics Applied With Excel*,
https://doi.org/10.1007/978-3-662-64319-8

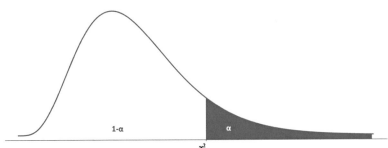

df	0.200	0.150	0.125	0.100	0.075	0.050	0.025	0.010	0.005
1	1.642	2.072	2.354	2.706	3.170	3.841	5.024	6.635	7.879
2	3.219	3.794	4.159	4.605	5.181	5.991	7.378	9.210	10.597
3	4.642	5.317	5.739	6.251	6.905	7.815	9.348	11.345	12.838
4	5.989	6.745	7.214	7.779	8.496	9.488	11.143	13.277	14.860
5	7.289	8.115	8.625	9.236	10.008	11.070	12.833	15.086	16.750
6	8.558	9.446	9.992	10.645	11.466	12.592	14.449	16.812	18.548
7	9.803	10.748	11.326	12.017	12.883	14.067	16.013	18.475	20.278
8	11.030	12.027	12.636	13.362	14.270	15.507	17.535	20.090	21.955
9	12.242	13.288	13.926	14.684	15.631	16.919	19.023	21.666	23.589
10	13.442	14.534	15.198	15.987	16.971	18.307	20.483	23.209	25.188
11	14.631	15.767	16.457	17.275	18.294	19.675	21.920	24.725	26.757
12	15.812	16.989	17.703	18.549	19.602	21.026	23.337	26.217	28.300
13	16.985	18.202	18.939	19.812	20.897	22.362	24.736	27.688	29.819
14	18.151	19.406	20.166	21.064	22.180	23.685	26.119	29.141	31.319
15	19.311	20.603	21.384	22.307	23.452	24.996	27.488	30.578	32.801
16	20.465	21.793	22.595	23.542	24.716	26.296	28.845	32.000	34.267
17	21.615	22.977	23.799	24.769	25.970	27.587	30.191	33.409	35.718
18	22.760	24.155	24.997	25.989	27.218	28.869	31.526	34.805	37.156
19	23.900	25.329	26.189	27.204	28.458	30.144	32.852	36.191	38.582
20	25.038	26.498	27.376	28.412	29.692	31.410	34.170	37.566	39.997
21	26.171	27.662	28.559	29.615	30.920	32.671	35.479	38.932	41.401
22	27.301	28.822	29.737	30.813	32.142	33.924	36.781	40.289	42.796
23	28.429	29.979	30.911	32.007	33.360	35.172	38.076	41.638	44.181
24	29.553	31.132	32.081	33.196	34.572	36.415	39.364	42.980	45.559
25	30.675	32.282	33.247	34.382	35.780	37.652	40.646	44.314	46.928
26	31.795	33.429	34.410	35.563	36.984	38.885	41.923	45.642	48.290
27	32.912	34.574	35.570	36.741	38.184	40.113	43.195	46.963	49.645
28	34.027	35.715	36.727	37.916	39.380	41.337	44.461	48.278	50.993
29	35.139	36.854	37.881	39.087	40.573	42.557	45.722	49.588	52.336
30	36.250	37.990	39.033	40.256	41.762	43.773	46.979	50.892	53.672
40	47.269	49.244	50.424	51.805	53.501	55.758	59.342	63.691	66.766
50	58.164	60.346	61.647	63.167	65.030	67.505	71.420	76.154	79.490
100	111.667	114.659	116.433	118.498	121.017	124.342	129.561	135.807	140.169
200	216.609	220.744	223.186	226.021	229.466	233.994	241.058	249.445	255.264

The F-Distribution

$\alpha = 10\%$

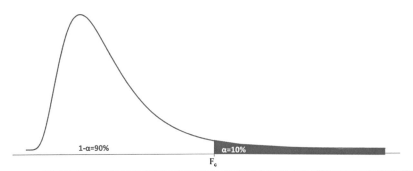

1-α=90%   α=10%

F_c

df$_2$	df$_1$													
	1	2	3	4	5	6	7	8	9	10	15	20	30	50
1	39.86	49.50	53.59	55.83	57.24	58.20	58.91	59.44	59.86	60.19	61.22	61.74	62.26	62.69
2	8.53	9.00	9.16	9.24	9.29	9.33	9.35	9.37	9.38	9.39	9.42	9.44	9.46	9.47
3	5.54	5.46	5.39	5.34	5.31	5.28	5.27	5.25	5.24	5.23	5.20	5.18	5.17	5.15
4	4.54	4.32	4.19	4.11	4.05	4.01	3.98	3.95	3.94	3.92	3.87	3.84	3.82	3.80
5	4.06	3.78	3.62	3.52	3.45	3.40	3.37	3.34	3.32	3.30	3.24	3.21	3.17	3.15
6	3.78	3.46	3.29	3.18	3.11	3.05	3.01	2.98	2.96	2.94	2.87	2.84	2.80	2.77
7	3.59	3.26	3.07	2.96	2.88	2.83	2.78	2.75	2.72	2.70	2.63	2.59	2.56	2.52
8	3.46	3.11	2.92	2.81	2.73	2.67	2.62	2.59	2.56	2.54	2.46	2.42	2.38	2.35
9	3.36	3.01	2.81	2.69	2.61	2.55	2.51	2.47	2.44	2.42	2.34	2.30	2.25	2.22
10	3.29	2.92	2.73	2.61	2.52	2.46	2.41	2.38	2.35	2.32	2.24	2.20	2.16	2.12
11	3.23	2.86	2.66	2.54	2.45	2.39	2.34	2.30	2.27	2.25	2.17	2.12	2.08	2.04
12	3.18	2.81	2.61	2.48	2.39	2.33	2.28	2.24	2.21	2.19	2.10	2.06	2.01	1.97
13	3.14	2.76	2.56	2.43	2.35	2.28	2.23	2.20	2.16	2.14	2.05	2.01	1.96	1.92
14	3.10	2.73	2.52	2.39	2.31	2.24	2.19	2.15	2.12	2.10	2.01	1.96	1.91	1.87
15	3.07	2.70	2.49	2.36	2.27	2.21	2.16	2.12	2.09	2.06	1.97	1.92	1.87	1.83
16	3.05	2.67	2.46	2.33	2.24	2.18	2.13	2.09	2.06	2.03	1.94	1.89	1.84	1.79
17	3.03	2.64	2.44	2.31	2.22	2.15	2.10	2.06	2.03	2.00	1.91	1.86	1.81	1.76
18	3.01	2.62	2.42	2.29	2.20	2.13	2.08	2.04	2.00	1.98	1.89	1.84	1.78	1.74
19	2.99	2.61	2.40	2.27	2.18	2.11	2.06	2.02	1.98	1.96	1.86	1.81	1.76	1.71
20	2.97	2.59	2.38	2.25	2.16	2.09	2.04	2.00	1.96	1.94	1.84	1.79	1.74	1.69
21	2.96	2.57	2.36	2.23	2.14	2.08	2.02	1.98	1.95	1.92	1.83	1.78	1.72	1.67
22	2.95	2.56	2.35	2.22	2.13	2.06	2.01	1.97	1.93	1.90	1.81	1.76	1.70	1.65
23	2.94	2.55	2.34	2.21	2.11	2.05	1.99	1.95	1.92	1.89	1.80	1.74	1.69	1.64
24	2.93	2.54	2.33	2.19	2.10	2.04	1.98	1.94	1.91	1.88	1.78	1.73	1.67	1.62
25	2.92	2.53	2.32	2.18	2.09	2.02	1.97	1.93	1.89	1.87	1.77	1.72	1.66	1.61
26	2.91	2.52	2.31	2.17	2.08	2.01	1.96	1.92	1.88	1.86	1.76	1.71	1.65	1.59
27	2.90	2.51	2.30	2.17	2.07	2.00	1.95	1.91	1.87	1.85	1.75	1.70	1.64	1.58
28	2.89	2.50	2.29	2.16	2.06	2.00	1.94	1.90	1.87	1.84	1.74	1.69	1.63	1.57
29	2.89	2.50	2.28	2.15	2.06	1.99	1.93	1.89	1.86	1.83	1.73	1.68	1.62	1.56
30	2.88	2.49	2.28	2.14	2.05	1.98	1.93	1.88	1.85	1.82	1.72	1.67	1.61	1.55
40	2.84	2.44	2.23	2.09	2.00	1.93	1.87	1.83	1.79	1.76	1.66	1.61	1.54	1.48
50	2.81	2.41	2.20	2.06	1.97	1.90	1.84	1.80	1.76	1.73	1.63	1.57	1.50	1.44
100	2.76	2.36	2.14	2.00	1.91	1.83	1.78	1.73	1.69	1.66	1.56	1.49	1.42	1.35
200	2.73	2.33	2.11	1.97	1.88	1.80	1.75	1.70	1.66	1.63	1.52	1.46	1.38	1.31

$\alpha = 5\%$

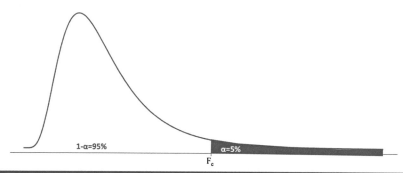

df$_2$	1	2	3	4	5	6	7	8	9	10	15	20	30	50
1	161.45	199.50	215.71	224.58	230.16	233.99	236.77	238.88	240.54	241.88	245.95	248.01	250.10	251.77
2	18.51	19.00	19.16	19.25	19.30	19.33	19.35	19.37	19.38	19.40	19.43	19.45	19.46	19.48
3	10.13	9.55	9.28	9.12	9.01	8.94	8.89	8.85	8.81	8.79	8.70	8.66	8.62	8.58
4	7.71	6.94	6.59	6.39	6.26	6.16	6.09	6.04	6.00	5.96	5.86	5.80	5.75	5.70
5	6.61	5.79	5.41	5.19	5.05	4.95	4.88	4.82	4.77	4.74	4.62	4.56	4.50	4.44
6	5.99	5.14	4.76	4.53	4.39	4.28	4.21	4.15	4.10	4.06	3.94	3.87	3.81	3.75
7	5.59	4.74	4.35	4.12	3.97	3.87	3.79	3.73	3.68	3.64	3.51	3.44	3.38	3.32
8	5.32	4.46	4.07	3.84	3.69	3.58	3.50	3.44	3.39	3.35	3.22	3.15	3.08	3.02
9	5.12	4.26	3.86	3.63	3.48	3.37	3.29	3.23	3.18	3.14	3.01	2.94	2.86	2.80
10	4.96	4.10	3.71	3.48	3.33	3.22	3.14	3.07	3.02	2.98	2.85	2.77	2.70	2.64
11	4.84	3.98	3.59	3.36	3.20	3.09	3.01	2.95	2.90	2.85	2.72	2.65	2.57	2.51
12	4.75	3.89	3.49	3.26	3.11	3.00	2.91	2.85	2.80	2.75	2.62	2.54	2.47	2.40
13	4.67	3.81	3.41	3.18	3.03	2.92	2.83	2.77	2.71	2.67	2.53	2.46	2.38	2.31
14	4.60	3.74	3.34	3.11	2.96	2.85	2.76	2.70	2.65	2.60	2.46	2.39	2.31	2.24
15	4.54	3.68	3.29	3.06	2.90	2.79	2.71	2.64	2.59	2.54	2.40	2.33	2.25	2.18
16	4.49	3.63	3.24	3.01	2.85	2.74	2.66	2.59	2.54	2.49	2.35	2.28	2.19	2.12
17	4.45	3.59	3.20	2.96	2.81	2.70	2.61	2.55	2.49	2.45	2.31	2.23	2.15	2.08
18	4.41	3.55	3.16	2.93	2.77	2.66	2.58	2.51	2.46	2.41	2.27	2.19	2.11	2.04
19	4.38	3.52	3.13	2.90	2.74	2.63	2.54	2.48	2.42	2.38	2.23	2.16	2.07	2.00
20	4.35	3.49	3.10	2.87	2.71	2.60	2.51	2.45	2.39	2.35	2.20	2.12	2.04	1.97
21	4.32	3.47	3.07	2.84	2.68	2.57	2.49	2.42	2.37	2.32	2.18	2.10	2.01	1.94
22	4.30	3.44	3.05	2.82	2.66	2.55	2.46	2.40	2.34	2.30	2.15	2.07	1.98	1.91
23	4.28	3.42	3.03	2.80	2.64	2.53	2.44	2.37	2.32	2.27	2.13	2.05	1.96	1.88
24	4.26	3.40	3.01	2.78	2.62	2.51	2.42	2.36	2.30	2.25	2.11	2.03	1.94	1.86
25	4.24	3.39	2.99	2.76	2.60	2.49	2.40	2.34	2.28	2.24	2.09	2.01	1.92	1.84
26	4.23	3.37	2.98	2.74	2.59	2.47	2.39	2.32	2.27	2.22	2.07	1.99	1.90	1.82
27	4.21	3.35	2.96	2.73	2.57	2.46	2.37	2.31	2.25	2.20	2.06	1.97	1.88	1.81
28	4.20	3.34	2.95	2.71	2.56	2.45	2.36	2.29	2.24	2.19	2.04	1.96	1.87	1.79
29	4.18	3.33	2.93	2.70	2.55	2.43	2.35	2.28	2.22	2.18	2.03	1.94	1.85	1.77
30	4.17	3.32	2.92	2.69	2.53	2.42	2.33	2.27	2.21	2.16	2.01	1.93	1.84	1.76
40	4.08	3.23	2.84	2.61	2.45	2.34	2.25	2.18	2.12	2.08	1.92	1.84	1.74	1.66
50	4.03	3.18	2.79	2.56	2.40	2.29	2.20	2.13	2.07	2.03	1.87	1.78	1.69	1.60
100	3.94	3.09	2.70	2.46	2.31	2.19	2.10	2.03	1.97	1.93	1.77	1.68	1.57	1.48
200	3.89	3.04	2.65	2.42	2.26	2.14	2.06	1.98	1.93	1.88	1.72	1.62	1.52	1.41

$\alpha = 1\%$

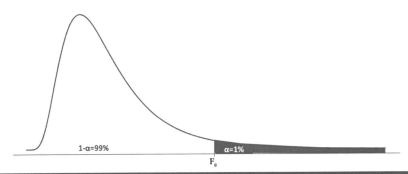

df$_2$	df$_1$													
	1	2	3	4	5	6	7	8	9	10	15	20	30	50
1	4052.2	4999.5	5403.4	5624.6	5763.6	5859.0	5928.4	5981.1	6022.5	6055.8	6157.3	6208.7	6260.6	6302.5
2	98.50	99.00	99.17	99.25	99.30	99.33	99.36	99.37	99.39	99.40	99.43	99.45	99.47	99.48
3	34.12	30.82	29.46	28.71	28.24	27.91	27.67	27.49	27.35	27.23	26.87	26.69	26.50	26.35
4	21.20	18.00	16.69	15.98	15.52	15.21	14.98	14.80	14.66	14.55	14.20	14.02	13.84	13.69
5	16.26	13.27	12.06	11.39	10.97	10.67	10.46	10.29	10.16	10.05	9.72	9.55	9.38	9.24
6	13.75	10.92	9.78	9.15	8.75	8.47	8.26	8.10	7.98	7.87	7.56	7.40	7.23	7.09
7	12.25	9.55	8.45	7.85	7.46	7.19	6.99	6.84	6.72	6.62	6.31	6.16	5.99	5.86
8	11.26	8.65	7.59	7.01	6.63	6.37	6.18	6.03	5.91	5.81	5.52	5.36	5.20	5.07
9	10.56	8.02	6.99	6.42	6.06	5.80	5.61	5.47	5.35	5.26	4.96	4.81	4.65	4.52
10	10.04	7.56	6.55	5.99	5.64	5.39	5.20	5.06	4.94	4.85	4.56	4.41	4.25	4.12
11	9.65	7.21	6.22	5.67	5.32	5.07	4.89	4.74	4.63	4.54	4.25	4.10	3.94	3.81
12	9.33	6.93	5.95	5.41	5.06	4.82	4.64	4.50	4.39	4.30	4.01	3.86	3.70	3.57
13	9.07	6.70	5.74	5.21	4.86	4.62	4.44	4.30	4.19	4.10	3.82	3.66	3.51	3.38
14	8.86	6.51	5.56	5.04	4.69	4.46	4.28	4.14	4.03	3.94	3.66	3.51	3.35	3.22
15	8.68	6.36	5.42	4.89	4.56	4.32	4.14	4.00	3.89	3.80	3.52	3.37	3.21	3.08
16	8.53	6.23	5.29	4.77	4.44	4.20	4.03	3.89	3.78	3.69	3.41	3.26	3.10	2.97
17	8.40	6.11	5.18	4.67	4.34	4.10	3.93	3.79	3.68	3.59	3.31	3.16	3.00	2.87
18	8.29	6.01	5.09	4.58	4.25	4.01	3.84	3.71	3.60	3.51	3.23	3.08	2.92	2.78
19	8.18	5.93	5.01	4.50	4.17	3.94	3.77	3.63	3.52	3.43	3.15	3.00	2.84	2.71
20	8.10	5.85	4.94	4.43	4.10	3.87	3.70	3.56	3.46	3.37	3.09	2.94	2.78	2.64
21	8.02	5.78	4.87	4.37	4.04	3.81	3.64	3.51	3.40	3.31	3.03	2.88	2.72	2.58
22	7.95	5.72	4.82	4.31	3.99	3.76	3.59	3.45	3.35	3.26	2.98	2.83	2.67	2.53
23	7.88	5.66	4.76	4.26	3.94	3.71	3.54	3.41	3.30	3.21	2.93	2.78	2.62	2.48
24	7.82	5.61	4.72	4.22	3.90	3.67	3.50	3.36	3.26	3.17	2.89	2.74	2.58	2.44
25	7.77	5.57	4.68	4.18	3.85	3.63	3.46	3.32	3.22	3.13	2.85	2.70	2.54	2.40
26	7.72	5.53	4.64	4.14	3.82	3.59	3.42	3.29	3.18	3.09	2.81	2.66	2.50	2.36
27	7.68	5.49	4.60	4.11	3.78	3.56	3.39	3.26	3.15	3.06	2.78	2.63	2.47	2.33
28	7.64	5.45	4.57	4.07	3.75	3.53	3.36	3.23	3.12	3.03	2.75	2.60	2.44	2.30
29	7.60	5.42	4.54	4.04	3.73	3.50	3.33	3.20	3.09	3.00	2.73	2.57	2.41	2.27
30	7.56	5.39	4.51	4.02	3.70	3.47	3.30	3.17	3.07	2.98	2.70	2.55	2.39	2.25
40	7.31	5.18	4.31	3.83	3.51	3.29	3.12	2.99	2.89	2.80	2.52	2.37	2.20	2.06
50	7.17	5.06	4.20	3.72	3.41	3.19	3.02	2.89	2.78	2.70	2.42	2.27	2.10	1.95
100	6.90	4.82	3.98	3.51	3.21	2.99	2.82	2.69	2.59	2.50	2.22	2.07	1.89	1.74
200	6.76	4.71	3.88	3.41	3.11	2.89	2.73	2.60	2.50	2.41	2.13	1.97	1.79	1.63

Printed in the United States
by Baker & Taylor Publisher Services